Macmillan Building and Surveying Series
Series Editor: IVOR H. SEELEY
    Emeritus Professor, The Nottingham Trent University

Civil Engineering Quantities

Ivor H. Seeley

BSc, MA, PhD, FRICS,
CEng, FICE, FCIOB, FIH

*Chartered Quantity Surveyor and Chartered Civil Engineer,
Emeritus Professor of Nottingham Trent University*

Fifth Edition

MACMILLAN

First published 1965 by
THE MACMILLAN PRESS LTD
Houndmills, Basingstoke, Hampshire RG21 2XS
and London
Companies and representatives
throughout the world

ISBN 0–333–58906–8 hardcover
ISBN 0–333–58907–6 paperback

A catalogue record for this book is available
from the British Library.

Second (SI) edition 1971
Third edition 1977
Fourth edition 1987
Fifth edition 1993
Reprinted 1994

Printed in Great Britain by
Mackays of Chatham PLC
Chatham, Kent

'Read not to contradict and confute;
nor to believe and take for granted;
not to find talk and discourse;
but to weigh and consider.'
                              Francis Bacon

Contents

x   Contents

Preface to Fifth Edition

This book has been updated and rewritten to take account of the changes introduced by the third edition of the *Civil Engineering Standard Method of Measurement* (CESMM3) and the relevant provisions of the sixth edition of the ICE Conditions of Contract. Steps have also been taken to extend the usefulness of the book.

The second edition to CESMM sought to secure a clearer, more comprehensive and increasingly rationalised approach to measurement. The continued use of method-related charges has been welcomed as their use on a wide scale has reduced Contractors' cash flow problems and assisted considerably with the pricing of variations. Not all costs are proportional to the quantities of permanent work on the majority of civil engineering contracts, and the use of method-related charges assists in the identification and assessment of such costs.

This edition examines the third edition to CESMM in detail and applies the Method to a wide range of constructional work by means of worked examples, including the new sections in classes Y and Z. The worked examples contain suitable drawings and the hand-written dimensions are accompanied by extensive explanatory notes describing the approach to measurement and the application of the general principles, work classifications and accompanying rules contained in CESMM3. The CESMM3 coding has been included in the worked examples for identification purposes.

The provisions in the ICE Conditions of Contract (Sixth Edition) appertaining to measurement and financial aspects of contracts are examined in some detail together with the more significant consequences.

Finally, the various bill preparation processes are described and illustrated, including the use of computers.

It is hoped that this new edition will serve the needs of practising civil engineers and quantity surveyors and students of these disciplines, in the measurement of this class of work, even more effectively than the previous editions.

Nottingham   IVOR H. SEELEY
Summer 1992

Acknowledgements

The author expresses his thanks to The Institution of Civil Engineers for kind permission to quote from the *Civil Engineering Standard Method of Measurement* (Third Edition) and the *General Conditions of Contract for use in connection with Works of Civil Engineering Construction* (Sixth Edition).

The Ellis School of Building and Surveying kindly gave consent for the use of some of the concepts contained in course material prepared by the author for the School in years past.

The author is deeply indebted to Dr Martin Barnes for his kind permission to refer to his authoritative work, *The CESMM3 Handbook* (Thomas Telford, London, 1992) which comprehensively amplifies the third edition of the *Civil Engineering Standard Method of Measurement*.

Grateful thanks are also due to the publishers for abundant help and consideration during the production of the book, and for consent to quote from *Civil Engineering Contract Administration and Control*.

Ronald Sears prepared the additional handwritten dimensions and amendments to the original sheets, which are of outstanding quality and which contribute so much to the character and usefulness of the book.

The cover photograph is Gibraltar Dockyard Refurbishment by courtesy of Shand Ltd, Shand House, Matlock, Derbyshire DE4 3AF

Subjects Illustrated by Examples

A list of abbreviations used in the book will be found in Appendix I.

1 Scope of Civil Engineering Works and Method of Measurement

It is considered desirable to begin by defining some of the terms that are extensively employed in the 'measurement' of civil engineering works, so that their meaning and purpose shall be generally understood. The term 'measurement' covers both (1) 'taking-off' dimensions by scaling or reading from drawings and entering them in a recognised form on specially ruled paper, called 'dimensions paper' (illustrated on page 73), and (2) the actual measurement of the work as executed on the site.

The term 'quantities' refers to the estimated amounts of civil engineering work required in each operation or activity, and together these items give the total requirements of the contract.

These quantities are set down in a standard form on 'billing paper', as illustrated on page 57, which has been suitably ruled in columns, so that each item of work may be conveniently detailed with a brief description of the work, the probable quantity involved and a reference or coding. The billing paper also contains columns in which the contractor tendering for the particular project enters the rates and prices for each item of work. These prices, when added together, give the 'Contract Price' or 'Tender Total'.

Recognised 'units of measurement' are detailed in the *Civil Engineering Standard Method of Measurement* (CESMM3), issued by the Institution of Civil Engineers (reference I at end of book). This covers the majority of items of civil engineering work that are normally encountered. Most items are measured in metres, and may be cubic, square or linear. Some items such as structural steelwork and metalwork and steel reinforcing rods or bars are measured by weight, in which case the tonne will be the appropriate unit of measurement.

The primary function of a 'bill of quantities' is to set down the various items of work in a logical sequence and recognised manner, so that they may be readily priced by contractors. The bill of quantities thus enables all contractors to tender on the same information. It also

1

provides a good basis for the valuation of 'variations' resulting from changes in design as the work proceeds.

A contractor will build up in detail a price for each item contained in the bill of quantities, allowing for the cost of the necessary labour, materials and possibly associated plant, together with the probable wastage on materials, and sometimes appropriate temporary work, establishment charges or oncosts and profit. It is most important that each billed item should be so worded that there is no doubt at all in the mind of a contractor as to the nature and extent of the item that he is pricing. This is assisted by the changed approach to measurement introduced by the second and third editions of CESMM, which ensures greater uniformity in the format and content of civil engineering bills of quantities. Contractors generally tender in keen competition with one another and this calls for very skilful pricing by contractors to secure contracts.

The subject of estimating for civil engineering contracts is outside the scope of this book, but detailed information on this subject can be found in McCaffer,[2] Spon[3] and Wessex.[4]

The bill of quantities backs up the contract documents, which normally consist of the conditions of contract, drawings and specification. The main function of the bill is to identify work and to enable prices to be entered against it.[5]

Development of Civil Engineering Codes of Measurement

The Institution of Civil Engineers published a report of a committee dealing with engineering quantities in 1933, and thus provided for the first time a standard procedure for drafting bills of quantities for civil engineering work. Prior to the introduction of this document there was no uniformity of practice in the measurement of civil engineering quantities, and engineers responsible for the preparation of civil engineering bills of quantities largely worked up their own systems of measurement as they thought fit. The order and nature of the billed items, the units of measurement and even the method of tabulating the information in specific columns — usually referred to as 'ruling' — which was adopted for the bills of quantities, varied considerably.

It will be appreciated that this lack of uniformity in the preparation of civil engineering bills of quantities made the task of civil engineering contractors in pricing them far more difficult than it is today, now that a more uniform method of measurement has been generally adopted.

In 1953, after much deliberation and consultation, a revised document, entitled the *Standard Method of Measurement of Civil Engineering Quantities*, was issued by the Institution of Civil Engineers, and this was reissued with slight amendments in 1963 and a metric addendum in 1968. This amended the previous (1933) report to meet the changing needs of civil engineers and contractors, and tied up with the provisions of the General Conditions of Contract for use in connection with Works of Civil Engineering Construction. Certain sections of the 1933 report were simplified, particularly those dealing with concrete and pipe lines. New sections covering site investigation and site clearance were added and provision was made for the measurement of prestressed concrete.

In 1967 the Construction Industry Research and Information Association (CIRIA) established a working party to identify research needs aimed at improving contract procedure. One of the projects that followed aimed at developing and testing an improved form of bill of quantities for civil engineering contracts, and the results were summarised in CIRIA report 34.[6] This study sought to define the needs of the industry and to propose means of making the information in the bill more useful, and so to reduce the high administrative cost of measurement. The investigations incorporated the use of experimental features of bills of quantities on live civil engineering contracts. The dominant conclusion in the report is that civil engineering bills of quantities, apart from scheduling the components of the contemplated work, should also contain charges related to the method and timing of the contractor's operations.

Following the publication of the CIRIA Report a steering committee was appointed by the Institution of Civil Engineers to undertake a detailed reappraisal of the civil engineering code of measurement. The steering committee spent five years formulating its proposals and throughout this period consulted extensively with the construction industry and other relevant bodies and persons. The resulting *Civil Engineering Standard Method of Measurement* was published in 1976.

The principal changes introduced by the CESMM were as follows.

(1) Greater standardisation of format, both in the component items and in the way they are described. A reduction of the previous variety that frequently arose from house styles and often led to unnecessary confusion to tendering contractors.
(2) It introduced various levels of classification or pigeon holes from which descriptions can be developed. There are also coding arrangements, which have no contractual significance, although they will assist where computers are used and form a useful basis for cost analysis.
(3) Use of method-related charges to represent more clearly site construction costs, such as the cost of setting up and operating plant, labour teams and the like. In these cases the best cost parameter is

not the finished physical work but what the contractor has to do on site.
(4) A large number of small changes to detailed rules of measurement, resulting in a tightening up and increased detailing and the removal of anomalies and some differences in interpretation. Cost is very much influenced by the location of work, and although it was not found practicable to frame rules to cover this, the engineer or surveyor preparing bills can reflect this in the way he prepares the information and describes the items. It was claimed that bills prepared in accordance with the CESMM would be easier to compile, be of greater use to the contractor, better reflect the costs involved and more effectively serve other purposes, such as programming, cost control and management.

The second edition of the *Civil Engineering Standard Method of Measurement* (CESMM2) was published in 1985, following two years of preparation work. The measurement notes were retitled 'rules' and expanded, rearranged and classified to make reference and interpretation easier. Additional items were inserted to keep pace with new technology, particularly in site investigation and geotechnical processes. A new section was included to cover sewer renovation. Bills produced under CESMM2 were more comprehensive and problem areas in the first edition suitably clarified. Furthermore, an attempt was made to secure greater compatability with building measurement practice, with further extension in SMM7.

The third edition of the *Civil Engineering Standard Method of Measurement* (CESMM3) updated the code, brought it into alignment with the sixth edition of the *ICE Conditions of Contract*, and new sections were introduced on water mains renovation and simple building works incidental to civil engineering works.

Scope of Civil Engineering Works

Before comparing the methods adopted for the measurement of civil engineering work with those used for building work, some consideration should be given to the nature and scope of civil engineering works, to appreciate fully and understand the need for a different and separate mode of measurement to operate in respect of these latter works. This comparison is included primarily for the use and guidance of quantity surveyors, many of whom are mainly concerned with the measurement of building works.

Civil engineering works encompass a wide range of different projects, some of which are of great magnitude. Vast cuttings and embankments; large mass and reinforced concrete structures, such as the frameworks of large buildings, reservoirs, sea walls, bridges and cooling towers for power stations; structural steel frameworks of large structures; piling for heavy foundations; jetties, wharves and dry docks; long pipelines, tunnels and railway trackwork, all form the subject-matter of civil engineering contracts.

Civil engineering work may also include structural engineering projects in reinforced concrete, steel, brick and timber, and public works engineering, such as roads, bridges, sewers, sewage treatment works, water mains, reservoirs, water towers, works of river and sea defence, refuse disposal plants, marinas and swimming pools, carried out on behalf of a variety of agencies as described in Public Works Engineering.[7]

These works require considerable skill, ingenuity and technical knowledge in both their design and construction. The use of new materials and techniques is continually changing the nature and methods of construction used in these projects, and the increasing size and complexity of these works demand a greater knowledge and skill for their measurement and valuation.

Some works involve elements of uncertainty, as for example the excavation work for extensive deep foundations or the laying of underground services under very variable site conditions. Many civil engineering projects are carried out on the banks of rivers or on the sea coast, and on low-lying marshy land, thus making the operations that are involved even more difficult and exacting. For these reasons it is essential that a code of measurement specially applicable to this class of work should be used.

Owing to the magnitude of most civil engineering works, it is advisable that the code of measurement adopted should be relatively simple, to avoid the separate measurement of labours and small items, some of which were dealt with separately when measuring building work, prior to the introduction of SMM7.[8] The term 'labours' referred to relatively small items of work, involving labour but no materials, such as forming fair battered angles to brickwork and holes in timber members. Furthermore, owing to the very nature of the works, there is a great deal more uncertainty than on building works, and the method of measurement needs to be more flexible to allow for variations in the methods of construction used and changes effected during the course of the constructional work made necessary by site conditions. The main function of a bill of quantities is to enable prices to be obtained for the project on a uniform basis and precise dimensions cannot always be prepared at the 'taking-off' stage. The quantities should always be as accurate as the drawings and other data permit but they can be adjusted following the measurement of the completed work on the site and the work, as executed, valued at billed or comparable rates, on what is often termed a 'measure and value contract'.

Extensive temporary works are likely to be required during the construction of civil engineering works and the contractor will need to cover the cost of these works in some part of the bill of quantities.

Comparison of Civil Engineering and Building Methods of Measurement

There are two separate and distinct practices of measurement operating for civil engineering and building works. There is, however, considerable common ground as regards the general approach, units of measurement employed and items of work that can be measured under both codes.

As previously stated, civil engineering work should be measured in accordance with the *Civil Engineering Standard Method of Measurement* prepared by the Institution of Civil Engineers and the Federation of Civil Engineering Contractors.[1] Building works are generally measured in accordance with the *Standard Method of Measurement of Building Works*, issued by the Royal Institution of Chartered Surveyors and the Building Employers Confederation.[8]

The details of building works are usually in a far more precise stage at the time of preparing the bill of quantities than is the case with civil engineering works. Furthermore, the building work normally covers more works sections, and is in consequence subject to more detailed measurement. In the absence of variations in design, most building work with the exception of sub-structural, drainage and external works, will not be subject to remeasurement and the contractor will be paid for the quantities of work incorporated in the bill of quantities.

In a building contract the bill of quantities will constitute a contract document, whereas in the majority of cases the specification will not. Consequently, the bill of quantities in a building contract will invariably be far more detailed with lengthier descriptions than that operating in civil engineering work, unless the components in the billed items, specification and drawings are coded in accordance with the recommendations for co-ordinated project information, which is the exception. The measurement of building work also involves a larger number of measured items, with the monetary rate entered against some of them being relatively small.

The Contractor, when pricing a civil engineering bill of quantities will need to refer constantly to the specification for detailed information on the billed items, and must generally include for any necessary temporary work and incidental labours. Furthermore, a much greater proportion of the work is likely to be below ground. There is, in consequence, a greater degree of risk in the pricing of civil engineering work.

In some of the larger civil engineering contracts there is also some building work. With these contracts the question often arises as to how the works as a whole are to be measured.

Take, for example, a large power station contract. The best procedure would appear to be to measure the main superstructure, the ancillary buildings and probably the chimneys in accordance with the *Standard Method of Measurement of Building Works*. The structural steel frameworks could be measured under either code of measurement. The remainder of the power station contract, comprising heavy foundations, piling, wharves and jetties, railway sidings, cooling towers, circulating water ducts, roads, sewers and water mains, are all essentially civil engineering work, and are best measured in accordance with the *Civil Engineering Standard Method of Measurement*.

In building work a larger number of items are measured separately, than is the case with civil engineering work. For instance, in building work, backfilling trenches, compacting trench bottoms and earthwork support are each measured separately, whereas in civil engineering work most of these items are included in the excavation rates.

CESMM2 introduced a different approach to measurement and pricing with three divisions of measurement in each work class and consequently greater uniformity in bill descriptions, and the use of method-related charges to permit the separation of items that are not proportional to the quantities of permanent work. SMM7[8] adopted a similar approach for building work and hence increased the amount of commonality in the approach adopted in the two measuring codes, which was also extended into some of the detailed measuring procedures, such as structural metalwork/steelwork.

Readers requiring further information on building measurement are referred to Seeley.[10,11]

2 Civil Engineering Contracts and Contract Documents

The first part of this chapter is concerned with the general characteristics of contracts and the remedies available when a contract is broken by a party to it. It gives the legal background to work under a contract and is required by many examining bodies. For more comprehensive and detailed information on the law of contracts, which is a most complicated subject, the reader might consult, for example, *Hudson's Building and Engineering Contracts*.[12]

The Nature and Form of Contracts

The law relating to civil engineering contracts is one aspect of the law relating to contract and tort or civil wrongs. It is therefore desirable to have some knowledge of the law relating to contracts generally before the main characteristics and requirements of civil engineering contracts are considered.

A simple contract consists of an agreement entered into by two or more parties, whereby one of the parties undertakes to do something in return for something to be undertaken by the other. A contract has been defined as an agreement which directly creates and contemplates an obligation. The word is derived from the Latin *contractum*, meaning drawn together.

We all enter into contracts almost every day for the supply of goods, transportation and similar services, and in all these instances we are quite willing to pay for the services we receive. Our needs in these cases are comparatively simple and we do not need to enter into lengthy or complicated negotiations and no written contract is normally executed. Nevertheless, each party to the contract has agreed to do something, and is liable for breach of contract if he fails to perform his part of the agreement.

In general, English law requires no special formalities in making contracts but, for various reasons, some contracts must be made in a particular form to be enforceable and, if they are not made in that special way, then they will be ineffective. Notable among these contracts are contracts for the sale and disposal of land, and 'land', for this purpose, includes anything built on the land, as, for example, roads, bridges and other structures.

Some contracts must be made 'under seal', for example, Deeds of Gift or any contract where 'consideration' is not present (consideration is defined later in the chapter). Some other contracts must be in writing, for example, that covering the Assignment of Copyright, where an Act of Parliament specifically states that writing is necessary. Contracts covering guarantee and land transactions may be made orally but will be unenforceable unless they are in writing, by virtue of the Law Reform (Enforcement of Contract) Act, 1954.

Since the passing of the Corporate Bodies Contracts Act, 1960, the contracts entered into by corporations, including local authorities, can be binding without being made under seal. The standing orders of most local authorities, however, will require major contracts to be made under seal, but the 1960 Act will avoid a repetition of the results of Wright v. Romford Corporation, where the local authority was able to avoid its responsibilities under a contract, merely because the contract had not been made under seal.

It is sufficient in order to create a legally binding contract, if the parties express their agreement and intention to enter into such a contract. If, however, there is no written agreement and a dispute arises in respect of the contract, then the Court that decides the dispute will need to ascertain the terms of the contract from the evidence given by the parties, before it can make a decision on the matters in dispute.

On the other hand if the contract terms are set out in writing in a document, which the parties subsequently sign, then both parties are bound by these terms even if they do not read them. Once a person has signed a document he is assumed to have read and approved its contents, and will not be able to argue that the document fails to set out correctly the obligations which he actually agreed to perform. Thus by setting down the terms of a contract in writing one secures the double advantage of affording evidence and avoiding disputes.

The law relating to contracts imposes on each party to a contract a legal obligation to perform or observe the terms of the contract, and gives to the other party the right to enforce the fulfilment of these terms or to claim 'damages' in respect of the loss sustained in consequence of the breach of contract.

Enforcement of Contracts

An agreement can only be enforced as a contract if:

(1) the agreement relates to the future conduct of one or more of the parties to the agreement;
(2) the parties to the agreement intend that their agreement shall be enforceable at law as a contract;
(3) it is possible to perform the contract without transgressing the law.

Validity of Contracts

The legal obligation to perform a contractual obligation only exists where the contract is valid. In order that the contract shall be valid the following conditions must operate.

(1) There must be an offer made by one person (the offeror), and the acceptance of that offer by another person (the offeree), to whom the offer was made. Furthermore, the offer must be definite, and made with the intention of entering into a binding contract. The acceptance of the offer must be absolute, be expressed by words or conduct, and be accepted in the manner prescribed or indicated by the person making the offer.

An offer is not binding until it is accepted and, prior to acceptance, the offer may come to an end by lapse of time, by revocation by the offeror or by rejection by the offeree, and in these cases there can be no acceptance unless the offer is first renewed.

(2) The contract must have 'form' or be supported by 'consideration'. The form consists of a 'deed', which is a written document, which is signed, sealed and delivered, and this type of contract is known as a 'formal contract' or contract made by deed.

If a contract is not made by deed, then it needs to be supported by consideration, in order to be valid, and this type of contract is known as a 'simple contract'. Consideration has been defined as some return, pecuniary or otherwise, made by the promisee in respect of the promise made to him.

(3) Every party to a contract must be legally capable of undertaking the obligations imposed by the contract. For instance, persons under 18 years of age may, in certain cases, avoid liability under contracts

into which they have entered. Similarly a corporation can only be a party to a contract if it is empowered by a statute or charter to enter into it.

(4) The consent of a party to a contract must be genuine. It must not be obtained by fraud, misrepresentation, duress, undue influence or mistake.

(5) The subject matter of the contract must be legal.

Remedies for Breach of Contract

Whenever a breach of contract occurs a right of action exists in the Courts to remedy the matter. The remedies generally available are as follows:

(1) Damages
(2) Order for payment of a debt
(3) Specific performance
(4) Injunction
(5) Rescission

Each of these remedies will now be considered further.

(1) *Damages*. In most cases a breach of contract gives rise to a right of action for damages. The 'damages' consist of a sum of money, which will, as far as it is practicable, place the aggrieved party in the same position as if the contract had been performed.

The parties to a contract, when entering into the agreement, may agree that a certain sum shall be payable if a breach occurs. This sum is usually known as 'liquidated damages', where it represents a genuine estimate of the loss that is likely to result from the breach of contract. Where, however, the agreed sum is in the nature of a punishment for the breach of contract, then the term 'penalty' is applied to it, and penalties are not normally recoverable in full.

For instance, in civil engineering contracts it is often stipulated that a fixed sum shall be paid per day or per week, if the contract extends beyond the agreed contract period. If this sum is reasonable it constitutes liquidated damages and, unlike a penalty, is recoverable in full.

(2) *Order for payment of a debt*. A debt is a liquidated or ascertained sum of money due from the debtor to the creditor and is recovered by an 'action of debt'.

(3) *Specific performance*. The term 'specific performance' refers to an order of the Court directing a party to a contract to perform his part of the agreement. It is now only applied by the Courts on rare occasions when damages would be an inadequate remedy, but specific

performance constitutes a fair and reasonable remedy and is capable of effective supervision by the Court. This remedy will not be given if it requires the constant supervision of the Court.

(4) *Injunction*. An injunction is an order of Court directing a person not to perform a specified act. For instance, if A had agreed not to carry out any further building operations on his land, for the benefit of B, who owns the adjoining land, and B subsequently observes A starting building operations, then B can apply to the Court for an injunction restraining A from building. Damages, in these circumstances, would not be an adequate remedy.

(5) *Rescission*. Rescission consists of an order of Court cancelling or setting aside a contract and results in setting the parties back in the position that they were in before the contract was made.

Main Characteristics of Civil Engineering Contracts

Most contracts entered into between civil engineering contractors and their employers are of the type known as 'entire' contracts. These are contracts in which the agreement is for specific works to be under-taken by the contractor and no payment is due until the work is complete.

In an entire contract, where the employer agrees to pay a certain sum in return for civil engineering work, which is to be executed by the contractor, the contractor is not entitled to any payment if he abandons the work prior to completion, and will be liable in damages for breach of contract. Where the work is abandoned at the request of the employer, or results from circumstances that were clearly foreseen when the contract was entered into and provided for in its terms, then the contractor will be entitled to payment on a *quantum meruit* basis, that is, he will be paid as much as he has earned.

It is, accordingly, in the employer's interest that all contracts for civil engineering work should be entire contracts to avoid the possibility of work being abandoned prior to completion. However, contractors are usually unwilling to enter into any contracts, other than the very smallest, unless provision is made for interim payments to them as the work proceeds. For this reason the standard form of civil engineering contract[13] provides for the issue of interim certificates at various stages of the works.

It is customary for the contract further to provide that a prescribed proportion of the sum due to the contractor on the issue of a certificate shall be withheld. This sum is known as 'retention money' and serves to insure the employer against any defects that may arise in the work. The contract does, however, remain an entire contract, and the contractor is not entitled to receive payment in full until the work is satisfactorily completed, the defects correction period expired and the defects correction certificate issued.

That works must be completed to the satisfaction of the employer, or his representative, does not give the employer the right to demand an unusually high standard of quality throughout the works, in the absence of a prior express agreement. Otherwise the employer might be able to postpone indefinitely his liability to pay for the works. The employer is normally only entitled to expect a standard of work that would be regarded as reasonable by competent persons with considerable experience in the class of work covered by the particular contract. The detailed requirements of the specification will have a considerable bearing on these matters.

The employer or promoter[14] of civil engineering works normally determines the conditions of contract, which define the obligations and performances to which the contractor will be subject. He often selects the contractor for the project by some form of competitive tendering and any contractor who submits a successful tender and subsequently enters into a contract is deemed in law to have voluntarily accepted the conditions of contract adopted by the employer.

The obligations that a contractor accepts when he submits a tender are determined by the form of the invitation to tender. In most cases the tender may be withdrawn at any time until it has been accepted and may, even then, be withdrawn if the acceptance is stated by the employer to be 'subject to formal contract' as is often the case.

The employer does not usually bind himself to accept the lowest or indeed any tender and this is often stated in the advertisement. A tender is, however, normally required to be a definite offer and acceptance of it gives rise legally to a binding contract.

Types of Contract Encountered in Civil Engineering Works

A variety of contractual arrangements are available and the engineer will often need to carefully select the form of contract which is best suited for the particular project. The employer is entitled to know the reasoning underlying the engineer's choice of contract.[15]

Contracts for the execution of civil engineering works may be broadly classified as follows.

(1) *Bill of quantities contracts.* This type of contract, which incorporates a bill of quantities priced by the contractor, is the most commonly used form of contract for works of civil engineering construction of all but the smallest in extent, where the quantities of the bulk of the work can be ascertained with reasonable accuracy before the work is

started. A bill of quantities is prepared in accordance with CESMM[1] giving, as accurately as possible, the quantities of each item of work to be executed, and the contractor enters a unit rate against each item of work. The extended totals are added together to give the tender total. This type of contract is a measure and value contract and the contract price is the sum to be ascertained and paid in accordance with the provisions contained in the ICE Conditions of Contract,[13] clauses 55, 56, 57, 58, 59, 60 and 61 as described later in the chapter.

The preparation of comprehensive bills of quantities for civil engineering works can have an important and far-reaching effect on the cost of the works to the employer. The contractor tendering for the specific contract is provided with a schedule giving brief identifying descriptions and estimated quantities of all the items of work involved. In the absence of such a bill of quantities, each contractor tendering will have to assess the amount of work involved and this will normally have to be undertaken in a very short period of time, in amongst other activities.

Under these circumstances a contractor, unless he is extremely short of work, is almost bound to price high to allow himself a sufficient margin of cover for any items that he may inadvertently have missed. Furthermore, there is no really satisfactory method of assessing the cost of variations and the contractor may feel obliged to make allowance for this factor also, when building up his contract price.

Bills of quantities greatly assist in keeping tender figures at a realistic level. They should be prepared, whenever possible, on all but the smallest civil engineering contracts, as they offer the following advantages.

(1) The contractor is paid for the actual amount of work done and thus limits the risk element borne by the contractor.
(2) While providing a fair basis for payment, there is the facility for dealing with altered work.
(3) Adjudication of tenders is relatively straightforward as all tenderers price on a comparable basis.
(4) The bill of quantities gives tendering contractors a clear conception of the work involved.
(5) Most contractors in the United Kingdom are thoroughly familiar with this type of contract and are thus better able to submit a realistic price for the work.[16]

(2) *Lump sum contracts*. In a lump sum contract the contractor undertakes to carry out certain specified works for a fixed sum of money. The nature and extent of the works are normally indicated on drawings and the nature of the materials and workmanship described

in a specification, but no bill of quantities is provided.

This form of contract is largely used in conjunction with works that are small in extent, and where the work is above ground and clearly visible, such as a road resurfacing contract.

It has, however, occasionally been used where the works required are uncertain in character, and by entering into a lump sum contract the employer hoped to place the onus on the contractor for deciding the full extent of the works and the responsibility for the payment of any additional costs, which could not be foreseen before the works were started. The employer would then pay a fixed sum for the works, regardless of their actual cost, and this constitutes an undesirable practice from the contractor's point of view.

(3) *Schedule contracts.* This type of contract may take one of two forms. The employer may supply a schedule of unit rates covering each item of work and ask the contractors, when tendering, to state a percentage above or below the given rates for which they would be prepared to execute the work. Alternatively, and as is more usual, the contractors may be requested to insert prices against each item of work, and a comparison of the prices entered will enable the most favourable offer to be ascertained. Approximate quantities are sometimes included to assist the contractors in pricing the schedules and the subsequent comparison of the tendered figures.

This type of contract is really only suitable for use with maintenance and similar contracts, where it is not possible to give realistic quantities of the work to be undertaken. In this form of contract it is extremely difficult to make a fair comparison between the figures submitted by the various contractors, particularly where approximate quantities are not inserted in the schedules, since there is no total figure available for comparison purposes and the unit rates may fluctuate extensively between the different tenderers.

Another advantage of the use of schedules is that they can be prepared quickly for projects of long duration. During the execution of the early stages of a project by a contractor selected from a schedule of adequately detailed rates, a bill of quantities can be prepared for the remainder of the work. This bill can be priced using the rates inserted in the original schedule by the contractor already employed on the site, or alternatively, competitive tenders can be obtained and, if appropriate, another contractor can carry out the later phases. For example, the substructure of a power station could be measured and valued in accordance with a schedule of rates, while the superstructure could be the subject of a bill of quantities contract. In like manner, approximate quantities could be provided for the first and possibly experimental section of a road or airport runway, with a more accurate bill of quantities prepared for the main scheme.[17]

A schedule contract also enables the contractor and the engineer to cooperate at the design stage in the development of new techniques in an effective and economical way. The schedule of rates should ideally be negotiated at an early stage to give the contractor and the engineer an opportunity to discuss the relationship of plant usage and site organisation to the design of the scheme.[17]

(4) *Cost reimbursement contracts.* In cost reimbursement contracts the employer pays to the contractor the actual cost of the work plus a management fee which will include the contractor's overhead charges, supervision costs and profit. The management fee may be calculated in one of four different ways which are now described.

(i) *Prime cost plus percentage contracts.* This type of contract provides for the management fee payable to the contractor to be calculated as a percentage of the actual or allowable total cost of the civil engineering work. It permits an early starting date, as the only matter requiring agreement between the employer and the contractor is the percentage to be applied in respect of the contractor's overheads and profit. It is accordingly relatively simple to operate and was used extensively during the Second World War for defence installations, and was subject to considerable abuse on occasions.

It is a generally unsatisfactory contractual arrangement as higher costs also entail higher fees and there is accordingly no incentive for efficiency and economy. The use of this form of contract should therefore be confined to situations where the full nature and extent of the work are uncertain and urgent completion of the project is required, resulting in a critical situation. Even then every care should be taken to safeguard the employer's interests by employing a reputable contractor and arranging effective supervision of the work. The main deficiency is that an unscrupulous contractor could increase his profit by delaying the completion of the works. No incentive exists for the contractor to complete the works as quickly as possible or to try to reduce costs. Furthermore, the fee will fluctuate proportionately to any prime cost fluctuations but these will not necessarily bear any relation to any changes in the actual costs of management.[18]

A typical percentage fee might contain an addition of 100 per cent on the actual cost of wages, fares and allowances paid by the contractor to the foremen, operatives and staff (other than clerical, administrative and visiting staff) for time spent wholly on the works, together with amounts paid in respect of such wages for national insurance, graduated pensions, holidays with pay, employer's liability and workmen's compensation insurance; an addition of 20 per cent on the actual cost of materials used upon the works after the deduction of all trade, cash and other discounts and rebates; an addition of 5 per cent on the actual cost of any sub-contractors' accounts in connection with the works and any payments made by the employer; and an addition of 10 per cent on the actual cost of any mechanical plant used on the site upon the works.[18]

(ii) *Prime cost plus fixed fee contracts.* In this form of contract the sum paid to the contractor will be the actual cost incurred in the execution of the works plus a fixed lump sum, which has previously been agreed upon and does not fluctuate with the final cost of the project. No real incentive exists for the contractor to secure efficient working arrangements on the site, although it is to his advantage to earn the fixed fee as quickly as possible and so release his resources for other work. This type of contract has advantages over the prime cost plus percentage contract from the employer's standpoint.

In order to establish a realistic figure for the fixed fee, it is necessary to be able to assess with reasonable accuracy the likely amount of the prime cost at the tender stage, otherwise the employer may have to revert to a prime cost plus percentage contract with its inherent disadvantages. It is advisable to prepare a document showing the estimated cost of the project in as much detail as possible so that the work is clearly defined and also the basis on which the fixed fee is calculated.

(iii) *Prime cost plus fluctuating fee contracts.* In this form of contract the contractor is paid the actual cost of the work plus a fee, with the amount of the fee being determined by reference to the allowable cost by some form of sliding scale. Thus, the lower the final cost of the works (prime cost), the greater will be the value of the fee that the contractor receives. An incentive then exists for the contractor to carry out the work as quickly and cheaply as possible, and it does constitute the most efficient of the three types of prime cost contract that have so far been described.

(iv) *Target cost contracts.* These are used on occasions to encourage the contractor to execute the work as cheaply as possible. A basic fee is generally quoted as a percentage of an agreed target estimate usually obtained from a priced bill of quantities. The target estimate may be adjusted for variations in quantity and design, and fluctuations in the cost of labour and materials and related matters. The actual fee paid to the contractor is determined by increasing or reducing the basic fee by an agreed percentage of the saving or excess between the actual cost and the adjusted target estimate. In some cases a bonus or penalty based on the completion time may also be applied.

Hence prime costs are recorded and a fee agreed for management services provided by the contractor as in the other forms of cost reimbursement contract. The actual amount paid to the contractor depends on the difference between the target price and the actual prime cost. In practice, various methods have been used for computing this sum. An alternative method that has been used is to pay the contractor the prime cost plus the agreed fee, and for the difference between target price and prime cost, whether a saving or an extra, to be shared between the employer and the contractor in agreed proportions. Yet another method is to pay either the target price or the prime cost plus the agreed fee whichever is the lower. This latter form of contract does, in fact, combine the characteristics of both the fixed price and cost reimbursement contracts.[18]

Fluctuations in fee due to difference between target and actual costs operate as a bonus to the contractor if his management is efficient, or as a penalty if it is inefficient. The benefits to be obtained by the employer from this contractual arrangement are mainly dependent on the target price being agreed at a realistic value, as there will be a great incentive for the contractor to increase the estimated price as much as possible in the first instance. It is essential that the employer obtains expert advice in evaluating this price. It may be negotiated with the contractor or established in competition. Target cost contracts should not be entered into lightly as they are expensive to manage, and require accurate management and careful costing on the employer's behalf.[19]

(5) *All-in contracts*. With this type of contract the employer, frequently using the services of an engineer, normally gives his requirements in broad outline to contractors, who are asked to submit full details of design, construction and cost, and probably including maintenance of the works for a limited period. This procedure has been used in the chemical and oil industries and for the design and construction of nuclear power stations. It is a contractual arrangement which sometimes finds favour with overseas employers. Although they may appear very attractive, they may give rise to many difficulties in implementation. As an ICE report[14] emphasises, they can prove particularly difficult to operate in times of high inflation and when the basic technology is changing rapidly. The term 'design and build' is now often used, particularly on building projects, to describe a procurement arrangement where a single organisation is responsible to the employer for both design and construction.

All-in contracts are sometimes referred to as package deals and, in practice, the arrangements may vary considerably, ranging from projects where the contractor uses his own professional design staff and undertakes both complete design and construction, to projects where the contractor, specialising in a particular form of construction, offers to provide a full service based on preliminary sketch plans provided by the employer's engineer. All-in contracts can be on a fixed price or cost reimbursement basis, competitive or negotiated, and can incorporate the management contracting system described later in the chapter. The employer may require the contractor to include the total financing of the project, in which case it is often referred to as a turnkey contract.[19] In yet another form of contract, the contractor designs, constructs, operates and maintains the project and receives the revenue for a prescribed concessionary period, often 10 to 30 years. An example of this type of arrangement is the Greater Manchester Metrolink. While another variant is the design and manage form of contract.

The selection of the contractor should be based on a brief of the employer's requirements. The brief should ideally be prepared by the employer's engineer and costed by him, so that contractors are tendering on a brief that is within the employer's budget. It is costly for contractors to tender for this type of contract in competition, as each contractor will have to produce a design to meet the brief and price for construction. Where this process is taken to excess at the tendering stage, it will result in an uneconomic use of resources. Hence, many contractors are not prepared to proceed beyond outline sketch design and an indicative price at the competitive tender stage. Furthermore, the evaluation and comparison of contractors' tenders is complicated as each contractor is likely to interpret the brief in a different way. Hence considerable adjustments may be needed to reduce them to a common basis for purposes of comparison.[19]

(6) *Negotiated contracts.* As a general rule, negotiation of a contract with a single contractor should take place only if it can be shown to result in positive advantages to the employer. There are a number of situations in which negotiations may be beneficial to the employer and some of the more common instances are now listed:

(1) The employer has a business relationship with the contractor.
(2) The employer finds it difficult, or even impossible, to finance the project in any other way.
(3) The employer has let a contract in competition, and then another contract of similar design comes on programme.
(4) In particular geographical areas where there may be only one contractor available to do the work.
(5) A certain contractor is the only one available with either the expertise or the special plant required to carry out the project.
(6) At times when the construction industry is grossly over-stretched and negotiation offers the best approach.
(7) Where a rapid start is required, as for example when the original contractor has gone into liquidation.[19]

The two principal methods of negotiation are:

(1) Using the competitive rates obtained for similar work undertaken under similar conditions in another contract; but there are many inherent problems in adjusting the existing rates to provide a basis for pricing the new work.
(2) An agreed assessment of the estimated cost, to which will be added an agreed percentage for head office overheads and profit, which can be subsequently documented in a normal bill of quantities.

There are certain essential features which are required if the negotiation is to proceed satisfactorily. These include equality of the negotiator for each party, parity of information, agreement as to the basis of negotiation and an approximate apportionment of cost between suitable heads, such as site management, contractor's own labour, direct materials, plant, contractor's own sub-contractors, nominated sub-contractors, nominated suppliers, provisional sums and contingencies, and head office overheads and profit.[19]

Advantages can accrue from a decision to select a contractor and to negotiate a contract sum with him. For instance, the contractor can be brought in at an early stage as a member of the design team, so that full advantage can be taken of his knowledge, experience and constructional resources. He can take an effective role in the planning process, which should help towards producing a better design solution at lower cost, and possibly with a shorter completion time. Further benefits may be secured if sub-contractors are brought in at the same time.

With the contractor appointed, agreement may be reached on the format of the bill of quantities which will be of the greatest use to the contractor in programming, progressing and cost controlling the project. Where bills are produced by computer, it is relatively easy to produce bills in any alternative form very quickly.[18]

There are, however, weaknesses inherent in the negotiating process, such as the length of time required for pricing and negotiation and that there is rarely any guarantee that a lower price will be obtained than by the normal competitive tendering procedure. It may be argued that the allowance for business risk is a matter of opinion, the anticipated profit is based on hope, and off-site overheads are dictated by the efficiency of the construction organisation. Hence the probability of negotiating a contractor's margin equal to, or less than, that prevailing in the competitive market is in all probability unlikely.[18]

(7) *Management contracts.* The management contract is a system whereby a main contractor is appointed, either by negotiation or in competition, to manage the construction of a project on behalf of the employer, but not to undertake the physical construction, and works

closely with the employer's professional advisor(s). All physical construction is undertaken by sub-contractors normally selected in competition. The management contractor provides common services to the sub-contractors such as welfare facilities, and plant and equipment that is not confined to one sub-contractor, and sufficient management both on and off the site to undertake the planning and management, co-ordination and control of the project. He is paid a fee for his services and in addition, the cost of his on-site management, common services and the cost of all work undertaken by sub-contractors.

The management contract, which emanated from the United States, is most appropriate to large, complex projects exceeding £20m in value, which exhibit particular problems that militate against the employment of fixed price contract procedures. Typical examples are:

(1) Projects for which complicated machinery is to be installed concurrently with the construction work.
(2) Projects for which the design process will of necessity continue throughout most of the construction period.
(3) Projects on which construction problems are such that it is necessary or desirable that the design and management team includes a suitably experienced contractor appointed on such a basis that his interests are largely synonymous with those of the employer's professional advisor(s).

The management contractor will be concerned with keeping the cost of the works within the project budget, made up of the estimated value of sub-contract packages, site management and other related costs, and termed the estimate of prime cost (EPC) for the project. He also reports to the employer on possible extras and deals with the sub-contractors in regard to such matters as claims for loss and expense and the settlement of accounts. The management contractor attends all design and progress meetings and it is good policy for a representative of the employer also to be in attendance. The management contractor will be able to report, among other matters, on the dates by which he will require design information and on any information that is already late.

An alternative to management contracting is construction management, where a construction manager is paid a fee to manage the project, but he does not enter into any contracts which are all with the employer. The manager carries virtually no risk and his sole allegiance is to the employer and the project. It is best suited to employers who regularly commission civil engineering projects, have considerable in-house expertise and wish to be involved in the day-to-day progress of the works.[19]

Form and Purpose of Contract Documents

Contract documents form the basis on which a civil engineering contractor will prepare his tender and carry out and complete the contract works. It is, accordingly, essential that the documents shall collectively detail all the requirements of the project in a comprehensive and unambiguous way. These documents also identify all the rights and duties of the main parties to the contract — the employer, engineer and contractor. Collectively they constitute a binding contract, whereby the contractor undertakes to construct works in accordance with the details supplied by the engineer and the employer agrees to pay the contractor in stages during the execution of the works in the manner prescribed in the contract.[15]

The contract documents normally used in connection with a civil engineering contract are as follows.

(1) Form of Agreement
(2) Conditions of Contract
(3) Specification
(4) Bill of Quantities
(5) Contract Drawings
(6) Form of Tender

Consideration will now be given to the nature and purpose of each of these documents.

(1) *Form of Agreement*

The Form of Agreement constitutes the formal agreement between the employer (promoter) and the contractor for the execution of the works, both permanent and temporary, in accordance with the other contract documents. The employer covenants to pay the contractor at the times and in the manner prescribed by the contract. The Form of Agreement incorporated in the ICE Conditions of Contract[13] is that generally used.

(2) *Conditions of Contract*

The Conditions of Contract define the terms under which the work is to be undertaken, the relationship between the employer (promoter), the engineer and the contractor, the powers of the engineer and the terms of payment. For many years it had been considered desirable to use a standard set of conditions which could, as far as practicable, be applicable to all civil engineering contracts. Any special conditions relating to an individual contract can be added to the general clauses.

In 1945 the Institution of Civil Engineers and the Federation of Civil Engineering Contractors issued a standard set of Conditions of Contract for use in connection with works of civil engineering construction. In the later editions of this document the Association of Consulting Engineers was concerned with its preparation, in addition to the other two bodies previously mentioned. The current edition is the sixth edition of the ICE Conditions issued in 1991.[13] In 1988, the ICE Conditions of Contract for Minor Works[20] was issued to cover smaller straightforward projects undertaken over short periods and the potential risks to both employer and contractor are small. Furthermore, other sets of conditions have been specially prepared to cover civil engineering works to be carried out overseas, of which the International (Civil) Conditions of Contract generally known as the FIDIC Conditions[21] are the most widely used, having evolved from the ICE Conditions, with provision for the insertion of the ruling language in which the contract is to be construed and interpreted. It was first introduced in 1957 and the fourth edition was published in 1987.

In 1991, the ICE published *The New Engineering Contract*[22] to provide maximum flexibility with a core contract and six main options encompassing a conventional contract with activity schedule; conventional contract with bill of quantities; target contract with activity schedule; target contract with bill of quantities; cost reimbursable contract; and management contract; with a variety of secondary options for use where necessary, to allow the employer to choose the version most appropriate to his needs.

For building work it is usual to make use of the standard conditions issued by the Joint Contracts Tribunal for the Standard Form of Building Contract, and generally referred to as the 'JCT Conditions'.[23] There are alternative forms for use with quantities, without quantities and with approximate quantities and there are, in addition, conditions specially devised for use on local authority contracts. In addition there are JCT Standard Forms of Building Contract for use with contractor's design, management contracts, intermediate form, minor building works, measured term contracts, and fixed fee form of prime cost.

Where the contract is of very limited extent and the use of the standard comprehensive set of conditions is not really justified, an abbreviated set of conditions, normally worked up from the appropriate set of standard conditions, can be adopted.

With certain specialised classes of civil engineering work the responsible authorities have seen fit to introduce a number of clauses, which modify or supplement the standard clauses of the 'ICE Conditions'. A typical example is the clauses prepared for use on power station contracts. The making of modifications to the ICE Conditions can lead to uncertainties and disputes which are better avoided, and they make the task of the contractor in preparing his tender more difficult. An ICE publication[14] emphasises the desirability of adding special conditions, where necessary, instead of making modifications to the standard conditions. There is a separate set of conditions for use on government contracts for building and civil engineering work.[24]

The ICE Conditions of Contract for use in connection with works of civil engineering construction are almost invariably included as one of the contract documents on a civil engineering contract. The principal clauses of the 'ICE Conditions',[13] covering the measurement and valuation of the works are clauses 51 and 52, dealing with alterations, additions and omissions, clauses 55, 56 and 57 covering measurement, clauses 58 and 59 relating to provisional and prime cost sums and nominated sub-contracts, and clauses 60 and 61, which are concerned with certificates and payment.

The content and effect of these clauses are now considered in some detail.

Alterations, Additions and Omissions

51(1) *Ordered Variations*

The Engineer has the power to vary any part of the Works, including temporary works, and ordered variations may include changes in the specified sequence, method or timing of construction and may be ordered during the defects correction period. The Engineer's powers do not, however, extend to variation of the terms and conditions of the Contract.[25] A wide variety of circumstances result in variations to the works, ranging from adverse weather conditions to delays occasioned by the finding of fossils on the site.

51(2) *Ordered Variations to be in Writing*

The Contractor should not make any variation unless he receives a written order or written confirmation of a verbal order from the Engineer. The Contractor may himself confirm in writing an oral order of the Engineer and the latter must give written contradiction forthwith, otherwise it is deemed to be a written order by the Engineer. Should the Engineer fail to confirm his oral instruction to vary the works, or contradicts a written confirmation already issued to the Contractor, then a dispute exists which is referrable to arbitration under clause 66.

51(3) *Variation not to Affect Contract*

No variation ordered in accordance with clauses 51(1) and 51(2) can invalidate a contract in any way and the value of the variations must be taken into account when determining the Contract Price.

51(4) *Changes in Quantities*

The Engineer is not required to issue a written order to cover changes in quantity of billed items, which do not arise from the issue of a variation order, as these will be adjusted automatically when measuring and valuing the work as executed.

52(1) *Valuation of Ordered Variations*

The Engineer is required to consult with the Contractor prior to ascertaining the value of variations ordered under clause 51. The value of ordered variations is to be determined in accordance with the following principles:

 (i) Where work is of a similar character and executed under similar conditions to work priced in the Bill of Quantities, it shall be valued at applicable rates and prices.
(ii) Where work is not of a similar character or is not executed under similar conditions, or is ordered during the Defects Correction Period, the rates and prices in the Bill of Quantities shall be used as a basis for valuation so far as may be reasonable, otherwise a fair valuation is to be made.

Normally evaluation will be secured by agreement between the Contractor and the Engineer; failing this the Engineer is empowered to determine the rate or price in accordance with the principles outlined and to notify the Contractor.

52(2) *Engineer to Fix Rates*

Variations may render some billed rates or prices unreasonable or inapplicable and either the Engineer or Contractor may give notice to the other that rates or prices should be varied. Such notice is required to be given before commencement of the varied work or as soon thereafter as practicable. The Engineer shall fix such rates as he considers reasonable and proper, presumably having regard to the component elements of the original rates or prices.[25]

52(3) *Daywork*

The Engineer may, if he considers it necessary or desirable, order in writing that any additional or substituted work shall be executed on a daywork basis. The Contractor is required to submit to the Engineer for his approval quotations for materials before ordering them. This procedure could result in delay to the works giving rise to a claim for extension of time under clause 44.[25] The payment of daywork is dealt with in clause 56(4).

52(4) *Notice of Claims*

If the Contractor intends to claim a higher rate or price than the one notified to him by the Engineer pursuant to clauses 52(1), 52(2) or 56(2), he is required to give written notice within 28 days of receiving the Engineer's notification. In the case of claims for additional payment, other than those arising under clauses 52(1), 52(2) or 56(2), the Contractor is required to give written notice to the Engineer of his intention to claim as soon as reasonably possible and, in any event, within 28 days after the occurrence of the event giving rise to the claim and to keep supporting records. The Engineer may without admitting liability instruct the Contractor to keep such records and to permit the Engineer to inspect them and receive copies if required.

After notification, the Contractor shall submit to the Engineer, as soon as is reasonably practicable, a first interim account giving detailed particulars of the claim including the amount claimed and the grounds of the claim. Thereafter, at the reasonable request of the Engineer, the Contractor shall provide an updated statement incorporating any further grounds on which it is based. If the Contractor fails to produce particulars within the prescribed time limits or in any other way prevents or prejudices the Engineer's investigations into the claim, the Engineer may restrict consideration to the particulars provided.

Approved payments against claims may be included in interim certificates issued pursuant to clause 60. To ensure that the Employer continuously has a reasonable forecast of the probable final value of the contract, the Contractor should notify his intention to claim additional payment at the earliest possible opportunity and to particularise his claim as soon as he is reasonably able to do so, quite apart from any specific provisions in the Conditions of Contract. However, claims may result from previously unforeseen contention arising during the finalising of the measured account.[25]

Measurement

55(1) *Quantities*

The quantities set out in the Bill of Quantities are estimated quantities of the work, but not necessarily actual or correct quantities, and may accordingly be varied.

55(2) *Correction of Errors*

Any error in description or omission from the Bill of Quantities shall be corrected by the Engineer and the value of the work actually executed is to be ascertained in accordance with clause 52. This provision does not extend to the rectification of errors, omissions or wrong estimates in the descriptions, rates and prices inserted in the Bill of Quantities by the Contractor.

This provision gives added importance to the use and effect of CESMM3, although a billed item that is clear and unambiguous will always take precedence over the Standard Method notwithstanding clause 57.[26]

56(1) *Measurement and Valuation*

Except as otherwise stated, the Engineer is required to ascertain and determine by measurement the value in accordance with the contract of work done under it.

56(2) *Increase or Decrease of Rate*

The Engineer is empowered, after consultation with the Contractor, to vary rates or prices where these have been rendered unreasonable or inapplicable as a result of fluctuation in quantities. Should the Contractor disagree with the Engineer's proposals, he should invoke the provisions of sub-clause 52(4)(a).

This sub-clause recognises that in tendering the Contractor has no choice but to accept the billed quantity, but at the same time, owing to the very nature of civil engineering work, the Engineer cannot be precise in calculating the quantities required. Adjustment of quantities can be either up or down. In general an increased quantity should result in more economic use of plant and a reduction of rate. However, it may not always be so, particularly if more distant tips have to be used or plant requirements are changed.[26]

56(3) *Attending for Measurement*

Where the Engineer requires any part of the Works to be measured he shall give reasonable advance notice in writing to the Contractor, who shall either attend or send a qualified agent to assist in making the measurement. If the Contractor fails to attend the measurement or send a representative, then the measurement made by the Engineer or his representative shall be taken as being the correct measurement. It is important that any delegation of the Engineer's duties under this clause and any limitations placed on them, shall be fully particularised and notified in writing to the Contractor.[25]

56(4) *Daywork*

The Contractor is required to supply to the Engineer such records, receipts and documents as proof of the amounts paid and/or costs incurred. These returns shall be in the form and delivered at the times directed by the Engineer and shall be agreed within a reasonable time. Where the Engineer so requires the Contractor shall submit quotations for materials for approval before ordering them.

   The Contractor shall be paid for daywork under the conditions and at the rates and prices contained in the Daywork Schedule included in the Bill of Quantities. In the absence of a Daywork Schedule the Contractor will be paid in accordance with the *Schedule of Dayworks carried out incidental to Contract Work*[27] current at the date the work is carried out. Daywork is paid at the schedule rates operative at the date of execution (not at tender date) and hence this work cannot be included in price fluctuation accounts.[26]

57 *Method of Measurement*

Except where expressly indicated in the Bill of Quantities, it shall be deemed to have been prepared in accordance with the *Civil Engineering Standard Method of Measurement, Third Edition (CESMM3)*.[1] CESMM3 will be inserted in the Appendix to the Form of Tender where it is used as the basis of measurement. It will be necessary to amend this clause if some other method of measurement, such as the building method, is adopted.

Provisional and Prime Cost Sums and Nominated Sub-Contracts

58(1) *Use of Provisional Sums*

Provisional sums can be included in the Contract for the execution of work, or supply of goods, materials or services by the Contractor or by a Nominated Sub-contractor.

58(2) *Use of Prime Cost Items*

Prime Cost (PC) Items can be used for the execution of work or for the supply of goods, materials and services by a sub-contractor nominated by the Engineer. Alternatively the Contractor may himself opt to perform these services and will be paid in accordance with his submitted quotation accepted by the Engineer or, in the absence of such a quotation, the value will be determined in accordance with clause 52.

58(3) *Design Requirements to be Expressly Stated*

If any Provisional Sum or Price Cost Item includes the provision of design or specification services in connection with Permanent Works or any equipment or plant to be provided, such as surveys or soil investigations, this shall be expressly stated in the Contract and included in any Nominated Sub-contract. The Contractor's obligation in respect of such services shall be limited to that which is expressly stated in accordance with this clause, since under clause 8 the Contractor is not responsible for the design of any Permanent Works.

59(1) *Nominated Sub-contractors — Objection to Nomination*

The Contractor is not obliged to enter into a sub-contract with any Nominated Sub-contractor against whom he raises reasonable objection or who declines to enter into a sub-contract with him containing the five provisions incorporated in this clause.

59(2) *Engineer's Action upon Objection to Nomination or upon Determination of Nominated Sub-contract*

This clause prescribes five alternative courses of action which the Engineer may take where the Contractor objects to the appointment of a Nominated Sub-contractor.

59(3) *Contractor Responsible for Nominated Sub-contractors*

This emphasises that the Contractor shall be responsible for the work executed or goods, materials or services supplied by a Nominated Sub-contractor.

59(4)(a) *Nominated Sub-contractor's Default*

Where the Contractor considers that an event has arisen which justifies the exercise of his right under the forfeiture clause to terminate the sub-contract or treat it as repudiated by the Nominated Sub-contractor, he shall immediately notify the Engineer in writing.

59(4)(b) *Termination of Sub-contract*

With the Engineer's written consent, the Contractor may give notice to the Nominated Sub-contractor expelling him from the Sub-contract works or rescinding the Sub-contract. If however the Engineer withholds consent, the Contractor is entitled to receive appropriate instructions under clause 13.

59(4)(c) *Engineer's Action upon Termination*

If the Nominated Sub-contractor is expelled from the Sub-contract works, the Engineer shall immediately take appropriate action under clause 59(2).

59(4)(d) *Delay and Extra Expense*

Where the Contractor, with the Engineer's consent, terminates the Nominated Sub-contract, he shall take all appropriate actions to recover all additional expenses that are incurred from the Sub-contractor or under the security provided pursuant to sub-clause 59(1)(d), including any additional expenses incurred by the Employer as a result of the termination.

59(4)(e) *Reimbursement of Contractor's Loss*

Where the Contractor fails to recover all his reasonable expenses of completing the Sub-contract works and all his proper additional expenses resulting from the termination, he will be reimbursed his unrecovered expenses by the Employer.

59(5) *Provisions for Payment*

For all work executed or goods, materials or services supplied by Nominated Sub-contractors, the following items shall be included in the Contract Price:

(a) The actual price paid or due to be paid by the Contractor under the terms of the sub-contract, unless resulting from a default of the Contractor, net of all trade and other discounts, rebates and allowances other than any discount obtainable by the Contractor for prompt payment.

(b) The sum (if any) provided in the Bill of Quantities for associated labours.

(c) In respect of all other charges and profit, a sum which is a percentage of the actual price paid or due to be paid, calculated at the rate set against the relevant item of prime cost in the Bill of Quantities or, where there is no such provision, at the rate inserted by the Contractor in the Appendix to the Form of Tender as the percentage for adjustment of sums set against Prime Cost Items.

59(6) *Production of Vouchers, etc.*

The Contractor shall where required by the Engineer produce all quotations, invoices, vouchers, sub-contract documents, accounts and receipts relating to expenditure on work carried out by all Nominated Sub-contractors.

59(7) *Payment to Nominated Sub-contractors*

Prior to issuing any certificate under clause 60, which includes payment to a Nominated Sub-contractor, the Engineer is entitled to receive proof from the Contractor that all payments under previous certificates, less retentions, have been paid or discharged.

In the event of the Contractor withholding such payments he should give written notice to the Engineer showing reasonable cause and proof that the Nominated Sub-contractor has been notified in writing.

The Employer is entitled to pay direct to a Nominated Sub-contractor, on an Engineer's certificate, any sum or sums, less retentions, withheld by the Contractor and to set-off any such payments against amounts due or which become due from the Employer to the Contractor.[25]

Nevertheless, it behoves the Engineer to exercise these provisions with discretion. It follows that payment must already have been made to the Contractor so that direct payment to the Nominated Sub-contractor is necessarily a double payment, and there must be a sufficiency of monies owing to the Contractor to provide set-off.[26]

Where the Engineer has certified and the Employer has made direct payment to the Nominated Sub-contractor, the Engineer shall in issuing any further certificate in favour of the Contractor deduct from the amount thereof the amount so paid but shall not withhold or delay the issue of the certificate itself when due to be issued under the terms of the contract.

Certificates and Payment

60(1) *Monthly Statements*

The Contractor is required to submit to the Engineer a monthly statement, in a form prescribed in the Specification where applicable, and
showing the estimated contract value of the Permanent Works up to
the end of the month, a list and the value of goods or materials
delivered to the Site but not yet incorporated in the Permanent Works,
a list and value of goods or materials listed in the Appendix to the Form
of Tender not yet delivered to the Site, but of which property is vested
in the Employer pursuant to clause 54, and the estimated amounts to
which the Contractor considers himself entitled covering such items as
Temporary Works, Contractor's Equipment and, with the operation of
CESMM3, this may include method-related charges. Although only a
statement is called for, it should be prepared in sufficient detail to
enable the Engineer or his Representative to check the submission. It is
good practice for the Contractor to submit a fully detailed statement
quarterly with summary statements in the intervening months.[28]
 No statement is, however, to be submitted where the Contractor
considers that the total estimated value will fall below the sum inserted
in the Appendix to the Form of Tender as being the Minimum Amount
of Interim Certificates under clause 60(2).
 Amounts payable in respect of Nominated Sub-contracts are to be
listed separately.

60(2) *Monthly Payments*

Following delivery by the Contractor to the Engineer or the Engineer's
Representative of the monthly statement under clause 60(1), the Engineer is required to certify and the Employer to pay the Contractor all
within 28 days of delivery of the Contractor's monthly statement. To
achieve this it will be necessary for the Contractor and Engineer to
work closely together both in the taking of measurements and agreement of rates.[26]
 Amounts certified by the Engineer for items listed in sub-clauses
60(1)(a) and (d) (Permanent Works, Temporary Works, Contractor's
Equipment and the like) will be subject to retention, while amounts
certified against items listed in sub-clauses 60(1)(b) and (c) (goods or
materials) will not be subject to retention. The Engineer is to show
separately amounts certified in respect of Nominated Sub-contracts.

60(3) *Minimum Amount of Certificate*

The Engineer is not bound to issue, but at the same time is not prevented from issuing, an interim certificate for a sum less than that stated in the Appendix to the Form of Tender, until the whole of Works have been certified as substantially complete.

60(4) *Final Account*

Within three months of the date of the Defects Correction Certificate the Contractor is required to submit to the Engineer a statement of final account and supporting documentation, showing in detail the value of work done in accordance with the contract together with all further sums the Contractor considers due to him under the Contract up to the date of the Defects Correction Certificate.

Within three months after receipt of the final account and all information reasonably required for its verification, the Engineer is required to issue a final certificate stating the amount which, in his opinion, is finally due under the Contract up to the date of the Defects Correction Certificate. The balance thus determined must be paid subject to clause 47 (liquidated damages) within 28 days of the date of the final certificate.

60(5) *Retention*

Retention as provided for in sub-clause 60(2)(a) shall be calculated as the difference between:

(a) an amount calculated at the rate indicated in and up to the limit set out in the Appendix to the Form of Tender upon the amount due to the Contractor based on sub-clauses 60(1)(a) and 60(1)(d) and
(b) any payment which shall have become due under clause 60(6).

60(6) *Payment of Retention*

(a) When a Certificate of Substantial Completion is issued relating to any section or part of the Works, the Contractor is entitled to one half of such proportion of the retention money deductible to date under sub-clause 60(5)(a) as the value of the section or part bears to the value of the whole of the Works completed to date as certified under sub-clause 60(2)(a), and this shall be added to the amount next certified as due to the Contractor under clause 60(2). However, the total of the amounts released shall not exceed one half of the limit of retention set out in the Appendix to the Form of Tender.

(b) When the Certificate of Substantial Completion is issued for the whole of the Works, the Contractor is entitled to one half of the retention money calculated in accordance with sub-clause 60(5)(a), and payments shall be made within 14 days of the issue of Certificates.

(c) Upon the expiration of the Defects Correction Period (or the latest period where there is more than one), the remainder of the retention money shall be paid to the Contractor within 14 days, irrespective of whether there are outstanding claims by the Contractor against the Employer. Although if outstanding work remains unexecuted by the Contractor, as referred to under clause 48 or any work pursuant to clauses 49 or 50, the Employer may withhold payment until the completion of such work, or an amount which, in the opinion of the Engineer, is equivalent to the cost of the work remaining to be executed.

60(7) *Interest on Overdue Payments*

Failure by the Engineer to certify or the Employer to pay in accordance with clauses 60(2), (4) or (6), or any finding of an arbitrator to such effect, renders the Employer liable to pay interest to the Contractor on any overdue payment calculated at a rate equivalent to 2 per cent per annum above the base lending rate of the bank specified in the Appendix to the Form of Tender. Where an arbitrator under clause 66 finds that an Engineer fails to certify sums due by a specific date, these will be regarded as overdue for payment.

60(8) *Correction and Withholding of Certificates*

The Engineer shall not in any interim certificate reduce sums previously certified to Nominated Sub-contractors, if the Contractor shall have already paid or be bound to pay the sums, notwithstanding his general entitlement to omit from any certificate the value of any work done, or goods, materials or services rendered with which he may be dissatisfied. If the Engineer in the final certificate shall reduce sums previously certified and paid to the Nominated Sub-contractors provision is made for the Employer to reimburse the Contractor. This provision necessitates careful certification of sums to Nominated Sub-contractors.

60(10) *Copy of Certificate for Contractor*

Every certificate issued by the Engineer under clause 60 shall be sent to the Employer and a copy to the Contractor, with a detailed explanation as appropriate.

61(1) *Defects Correction Certificate*

The Defects Correction Certificate, certifying the satisfactory construction, completion and maintenance of the Works, cannot be issued until the expiration of the Defects Correction Period (or the latest period where there is more than one) or until such later time as any works required under clauses 48, 49 and 50 have been completed to the Engineer's satisfaction.

The release of the second half of retention is to be made 14 days after the expiration of the Defects Correction Period but is not dependent on the issue of a Defects Correction Certificate. As noted previously, a deduction may be made from such retention money equivalent to the cost of any work ordered and outstanding under clauses 48, 49 and 50.

The Engineer is required to send a copy of the Defects Correction Certificate to the Contractor when it is issued to the Employer.

61(2) *Unfulfilled Obligations*

The issue of the Defects Correction Certificate shall not relieve the Contractor or the Employer of their obligations under the Contract.

Contract Price Fluctuations

The contract price fluctuations clause is optional and is normally included in contracts of over two years duration. It consists of a cost indices system of variation of price which superseded the former laborious procedure of calculating the price fluctuations from wage sheets and invoices.

There is a separate clause (FSS) covering contract price fluctuations for use in those cases where it is required to make special provision for adjustment of price fluctuations in respect of fabricated structural steelwork only or predominantly fabricated structural steelwork together with a negligible amount of civil engineering work.

The amount to be added to, or deducted from, the contract price is the net amount of the increase or decrease in cost to the contractor in carrying out the works. The index figures are compiled by the Department of the Environment[29] and comprise:

(1) the index of the cost of labour in civil engineering construction;
(2) the index of the cost of providing and maintaining constructional plant and equipment; and

(3) the indices of constructional materials prices in respect of aggregates, bricks and clay products generally, cements, cast iron products, coated roadstone for road pavements and bituminous products generally, fuel for plant to which DERV fuel index will be applied, fuel for plant to which gas oil index will be applied, timber generally, reinforcement, other metal sections, fabricated structural steel, and labour and supervision in fabricating and erecting steelwork.

The base index figure is the appropriate final index figure applicable to the date 42 days prior to the date for the return of tenders, while the current index figure applies to the completion date or the last day of the period to which the certificate relates, whichever is the earliest.

The fluctuations apply to the effective value, as and when included in the monthly statements by the contractor and certified by the engineer and are subject to retention in accordance with clause 60(4). Materials on site are not included in the effective value. Dayworks or nominated sub-contractors' work are only excluded from the effective value if based on actual cost or current prices.

The increase or decrease in the amounts otherwise payable under clause 60 is calculated by multiplying the effective value by a price fluctuation factor, which is the net sum of the products obtained by multiplying each of the proportions inserted by the contractor against labour, plant and materials in sub-clauses 4(a), (b) and (c), by a fraction, the numerator of which is the relevant current index figure minus the relevant base index figure, and the denominator of which is the relevant base index figure. Provisional index figures used in the adjustment of interim certificates shall be subsequently recalculated on the basis of the corresponding final index figures.

The simplified formula results in a degree of approximation due to:

(1) the extent to which the proportional factors inserted by the contractor vary from the operative figures on the site;
(2) the extent to which the factors and weightings forming the basis for each index vary from those appertaining to the particular contract; and
(3) the pattern of interim payments which may vary significantly from the pattern of costs incurred.[28]

Form and Purpose of Contract Documents (continued)

(c) *Specification*

The specification amplifies the information given in the contract drawings and the bill of quantities. It describes in detail the work to be executed under the contract and the nature and quality of the materials, components and workmanship. It gives details of any special responsibilities to be borne by the contractor, apart from those covered by the general conditions of contract. It may also contain clauses specifying the order in which the various sections of the work are to be carried out, the methods to be adopted in the execution of the work, and details of any special facilities that are to be afforded to other contractors.

Civil Engineering Procedure[14] issued by the Institution of Civil Engineers recommends that the specification should also require tenderers to submit a programme and a description of proposed methods and temporary works with their tenders. Care is needed when drafting a specification to avoid any conflict with provisions in the conditions of contract or bill of quantities.

The specification will always constitute a contract document in civil engineering contracts, while in the case of building contracts, under the JCT form of contract, it will only be a contract document, if there is no bill of quantities or when it is specifically made a contract document in the particular contract.

A Sub-committee of the Institution of Civil Engineers in a report entitled *The Contract System in Civil Engineering*, issued in 1946, drew attention to the desirability of standardising specifications, particularly with regard to materials, where there had been wide variations in the descriptions used. The use of British Standards helps considerably in this respect, ensuring the use of good quality materials, complying with the latest requirements prepared by expert technical committees representing the user, producer, research and other interests. Their use also simplifies the work of the engineer, since in most cases he no longer needs to draft clauses specifying in detail the materials to be used. In time, British Standards may be replaced by Eurocodes.

It is, however, most important that any references to British Standards should include the appropriate class or type of material required, where a number of classes or types are given in the British Standard; for example, clauses sometimes appear in specifications relating to 'first quality' and 'second quality' clay pipes complying with BS 65, whereas the classes of pipe recognised by that standard are 'British Standard', 'British Standard Surface Water' (SW) and 'British Standard Extra Chemically Resistant' (ECR).

An excellent arrangement for a specification covering civil engineering works is to start with any special conditions relating to the contract and the extent of the contract; then to follow with a list of contract drawings, details of the programme, description of access to the site, supply of electricity and water, offices and mess facilities, and statements regarding suspension of work during frost and bad weather, damage to existing services, details of borings, water levels and similar clauses.

This section could conveniently be followed by detailed clauses covering the various sections of the work, starting with materials in each case and then proceeding with workmanship and other clauses.

There is a considerable difference in the method of preparing specifications and bills of quantities for civil engineering work as compared with building work.

The civil engineering practice is to use brief descriptions in the items in the bill of quantities and to give more comprehensive and detailed information concerning the materials, components and workmanship, in the specification, which is also a contract document. With building contracts the billed item descriptions are more lengthy and preamble clauses at the head of each work section bill frequently take the place of the specification, which would not in any case be a contract document, where quantities form part of the contract.

The contractor tendering for a civil engineering project must therefore refer in many instances to the specification for the details he needs on which to build up his contract rates, while on a building contract, most of the information will be contained in the one document, that is, the bill of quantities. Once the contract is under way, the civil engineering method has much in its favour with a good comprehensive specification as a separate and strictly enforceable document. Guidance on the preparation of civil engineering specifications, accompanied by extensive examples, is contained in *Civil Engineering Specification*.[30]

(d) *Bill of Quantities*

The bill of quantities consists of a schedule of the items of work to be carried out under the contract with quantities entered against each item, the quantities being prepared in accordance with the *Civil Engineering Standard Method of Measurement* (CESMM3). Owing to the small scale of many of the drawings, the large extent of the works and the uncertainties resulting from difficult site conditions, the quantities inserted in a bill are often approximate. Nevertheless, the quantities should be as accurate as the information available allows and the descriptions accompanying each item must clearly identify the work involved.

The unit rates entered by the contractor against each item in the bill of quantities normally include all overhead charges and profit, but subject to the approach adopted in pricing method-related charges and the adjustment item in the grand summary. The contract usually makes provision for the quantities to be varied, and it is therefore highly desirable that separate items should be incorporated as method-related charges against which the contractor may enter the cost of meeting various contingent liabilities under the contract, such as special temporary works, and this aspect will be dealt with in more detail in chapter 4. The distribution of these liabilities over the measured items in the bill of quantities may make for difficulties in the event of any variations arising to the contract.

Provision is often made for the execution of certain work at daywork rates in a civil engineering bill of quantities.

One of the primary functions of a civil engineering bill of quantities is to provide a basis on which tenders can be obtained, and, when these are priced, they afford a means of comparing the various tenders received. After the contract has been signed, the rates in the priced bill of quantities can be used to assess the value of the work as executed.

(e) *Contract Drawings*

The contract drawings depict the details and scope of the works to be executed under the contract. They must be prepared in sufficient detail to enable the contractor to satisfactorily price the bill of quantities.

All available information as to the topography of the site and the nature of the soil and groundwater should be made accessible to all contractors tendering for the project. The contract drawings will be subsequently used when executing the works and may well be supplemented by further detailed drawings as the work proceeds.

Existing and proposed work should be clearly distinguished on the drawings and full descriptions and explanatory notes should be entered on them. The more explicit the drawings, the less likelihood will there be of disputes subsequently arising concerning the character or extent of the works. Ample figured dimensions should be inserted on drawings to ensure maximum accuracy in taking off quantities and in setting out constructional work on the site. Drawings may incorporate or be accompanied by schedules, such as those recording details of steel reinforcement and manholes.

For example, a contract encompassing a sewage treatment works, sewers and other associated work could be depicted on the following drawings: layout of sewage treatment works; working drawings of each section of the works, such as siteworks, inlet works, primary tanks, aeration tanks, secondary tanks, pumphouses, and culverts and pipework to a scale of 1:50 or 1:100; layout of the sewers and manholes often on a 1:2000 plan; longitudinal sections of sewers often to an exaggerated combined scale of 1:2000 horizontally and

1:500 vertically; manhole details to a scale of 1:50; and pumphouse details, where appropriate, to 1:50 scale.

(f) *Form of Tender*

The Form of Tender constitutes a formal offer to construct and complete the contract works in accordance with the various contract documents for the tender sum. It usually incorporates the contract period within which the contractor is to complete the works. Normally the tenderer submits a tender complying fully with the specification, but in some instances he is permitted to offer alternative forms of construction. The employer's written acceptance of the offer is binding, pending the completion of the agreement.

The form of tender now largely used for civil engineering contracts is the form incorporated in the ICE Conditions of Contract.[13] This form of tender provides for a 'bond' often amounting to 10 per cent of the tender sum. The contractor may be required to enter into a bond, whereby he provides a surety which is often a bank or insurance company who are prepared to pay up to, say, 10 per cent of the contract sum if the contract is not carried out satisfactorily.

The appendix to the form of tender is in two parts: part 1 being completed prior to the invitation of tenders and part 2 to be completed by the contractor. Part 1 to the appendix includes the defects correction period, the amount of the bond (if required), minimum amount of third party insurance, works commencement date, time for completion, amount of liquidated damages, vesting of materials not on site, method of measurement used, percentage of value of goods and materials to be included in interim certificates, minimum amount of interim certificates, rate and limit of retention and bank whose lending rate is to be used.

While part 2 incorporates insurance aspects, time for completion (if not included in part 1), vesting of materials not on site where at the option of the contractor, and percentage(s) for adjustment of PC sums.

The minimum amount of insurance is frequently calculated at 25 per cent of the tender total or £500 000, whichever is the greater. The time for completion is usually inserted by the engineer but is sometimes left for the contractor to insert. The defects correction period is usually 12 months. The percentage of the value of goods and materials to be included in interim certificates is often 80 per cent, while the minimum amount of interim certificates may be calculated at about one-half of the engineer's estimated average monthly value of the contract.

The Form of Tender is normally accompanied by Instructions to Tenderers, which aim to assist tenderers in the preparation of their tenders, and to ensure that they are presented in the form required by the Employer and the Engineer.

Invitation to Tender

The inviting of tenders for civil engineering works is usually performed by one of three methods

(1) by advertising for competitive tenders
(2) by inviting tenders from selected contractors
(3) by negotiating a contract with a selected contractor.

Advertisement for competitive tenders (open tendering) offers the most satisfactory method in some instances, since it ensures maximum competition. There is, however, the grave disadvantage that tenders may be received from firms who have neither the necessary financial resources nor adequate technical knowledge and experience of the class of work involved. Public authorities sometimes invite tenders in this way, although it results in more abortive tendering and waste of resources.

The invitation of tenders from a selected list of contractors (selective tendering) is the best procedure, offering maximum efficiency and economic advantage. It is particularly advisable when the works involved are of great magnitude or are highly complex in character, such as the construction of large power stations and harbour works. It is feasible to prepare selected lists, subject to periodic review, covering different value ranges and categories of works, ensuring the selection of contractors of established skill, integrity, responsibility and proven competence for work of the character and size contemplated.

Negotiation of a tender with a selected contractor is usually only advisable in special circumstances, as for instance when the contractor is already engaged on the same site, where space is very restricted, and is executing another contract there. This procedure might also be usefully adopted when it is required to make an early start with the work or where the contractor in question has exceptional experience of the type of work covered by the particular contract.

Every case should be considered on its merits when deciding the method to be used for the invitation of tenders. More detailed information on contractor selection can be found in *Civil Engineering Contract Administration and Control*[15] by the same author.

3 General Arrangement and Contents of Civil Engineering Bills of Quantities

The *Civil Engineering Standard Method of Measurement* Third Edition (CESMM3)[1] defines a 'Bill of Quantities' as a list of items giving brief identifying descriptions and estimated quantities of the work comprised in a Contract (1.7). (Perhaps the word 'concise' would be better substituted for 'brief'.) All references in brackets refer to paragraphs in the *Civil Engineering Standard Method of Measurement* (paragraphs being the terminology used in CESMM3).

A civil engineering bill is not intended to describe fully the nature and extent of the work in a contract, and in this respect differs fundamentally from a bill for building work. The descriptions of civil engineering billed items merely *identify* the work and the estimator pricing the bill will need to obtain most of the information he requires for estimating from the Drawings and Specification. He will use the Bill of Quantities to obtain information on estimated quantities and as the means of submitting prices to the Employer.

Barnes[5] has described how the need for civil engineering contract financial control arises from the difficulty experienced by the Employer in clearly identifying to the Contractor, his exact requirements and the difficulty of the Contractor in assessing accurately the probable cost of the work. To achieve effective control a Bill of Quantities must be prepared with the object of limiting these difficulties as far as is practicable.

Definitions

A number of definitions are contained in Section 1 of CESMM3[1] and the more important are now stated and examined.

The term *work* (1.5) differs from *Works* in the ICE Conditions of Contract,[13] having a wider coverage to embrace everything that the Contractor has to do and includes all his liabilities, obligations and risks.

The expression *expressly required* (1.6) is used extensively in

42

CESMM3,[1] and appears in Bills of Quantities to indicate that specific requirements will be shown on the Drawings, described in the Specification or ordered by the Engineer. Typical examples of its use in CESMM3[1] are rule A2 of class B relating to trial pits and trenches involving hand excavation, rule M13 of class E relating to the double handling of excavated material and rule M15 of class E covering timber or metal supports left in excavations. In these cases, the work will only be measured and paid for when ordered by the Engineer. It is important that agreement should be reached between the Contractor and the Engineer's Representative as to the extent of the express requirement, and that this shall be recorded before work is started.

Daywork (1.8) is the method of valuing work on the basis of time spent by operatives, materials used and plant employed, with an allowance to cover oncosts and profit. This basis of valuation is considered in more detail later in this chapter and may be used where billed or adjusted rates would be inappropriate.

The CESMM3[1] contains four surface definitions to be used in excavation and associated work, with the object of avoiding uncertainty in the excavation level descriptions. *Original Surface* (1.10) denotes the original surface of the ground before any work is carried out. As work proceeds the Contractor is likely to excavate to lower surfaces, eventually producing the *Final Surface* (1.11) as shown on the Drawings, and being the surface that will normally receive the permanent work. There may also be intermediate stages, as for instance first to excavate to formation or base level of a road, below which excavation is needed over certain areas for catchpits and other features. The reduced or formation level constitutes the *Excavated Surface* (1.13) for the main excavation and the *Commencing Surface* (1.12) for the lower excavation, while the lowest excavated surface (base of catchpit) constitutes both the *Final Surface* (1.11) and the *Excavated Surface* (1.13) for the lower pocket of excavation. The substitution of the term 'surface' for 'level' stems from the fact that all excavated and original surfaces are not necessarily level.

Paragraph 1.14 enables ranges of dimensions in billed descriptions to be reduced in length and ensures uniformity of presentation and interpretation. For example, general excavation could be billed with a maximum depth range of 2–5 m, signifying that the excavation is to depths exceeding 2 m but not exceeding 5 m.

General Principles

The CESMM3[1] is concerned with the measurement of civil engineering work, but where some complex building, mechanical engineering, electrical engineering, or other work is included in a civil engineering contract, then this work should be adequately itemised and described and the method of measurement stated in the Preamble to the Bill of Quantities (2.2 and 5.4). Large civil engineering contracts often include some items of relatively simple building work, such as the superstructure to a pumphouse on a sewage treatment works. In these circumstances, the work will be classified as simple building works incidental to civil engineering works and measured in accordance with class Z of CESMM3,[1] as illustrated in example XI.

Maintenance and repair work, and alterations to existing work are not generally mentioned in CESMM3.[1] As a general rule itemisation and description of such work should follow the principles prescribed for the relevant class of new work. Reference to extraction or removal will need to be given in billed descriptions or headings.[5] The principal exceptions relate to sewer and water main renovation and ancillary works which are covered in class Y and illustrated in examples XIX and XX.

Occasionally it will be necessary to deal with components which are not covered in CESMM3[1] and non-standard items should be inserted in the bill which adequately describe the work and its location. Paragraph 2.5 emphasises the principle of providing for the inclusion of possible cost differentials arising from changes in location or any other aspects, by giving the Contractor the opportunity to make allowance for these cost differences in his rates and prices.

Barnes[5] illustrates this concept by reference to the cost-significant factors involved in pipe trench excavation, such as the practicability of battering the sides of trenches, the existence or otherwise of boulders, adequacy of working space and related matters. For this reason lengths of pipelaying are billed separately with locations indicated by reference to the Drawings as rule A1 of class I. The probable impact of construction costs must be considered and the work suitably subdivided in the bill to indicate the likely influence of location on cost.

In some instances CESMM3[1] states that certain procedures *may* or *should* be employed as against the more positive direction used in other parts of the Method incorporating the word *shall*. In the former instances there is no infringement of the CESMM if the procedure is not followed and this is intentional. For example, paragraph 4.3 indicates that code numbers may be used to number items in Bills of Quantities and 5.22 recommends that Bills should be set out on A4 size paper — both relate to bill layout and arrangement and have no contractual significance. Paragraph 5.8 suggests that priceable items in the Bill of Quantities may be arranged into numbered parts, while 5.10 recommends that the inclusion of further itemisation and additional description may be provided to incorporate factors which could give rise to special methods of construction or cost considerations. In the latter instances the person preparing the bill has to exercise his judgement on likely cost significant factors, and it would be quite unrealistic to adopt more positive terminology, which could entitle the Contractor to a Bill amendment.

Application of the Work Classification

The Work Classification provides the basic framework of CESMM3[1] constituting, as it does, a list of the commonly occurring components of civil engineering work. It will assist with Contractor's cost control, recording of prices as a basis for pre-contract estimating, computer processing and specification preparation.

There are 26 main classes of work, with each class made up of three divisions, which classify work at successive levels of detail. Each division contains a list of up to eight descriptive features of work. Each item description in the Bill of Quantities will incorporate one feature from each division of the relevant class (3.1). For example, Class H (precast concrete) contains three divisions — the first classifies different types of precast concrete units (beams, columns, slabs and the like), the second classifies the different units by their dimensions (lengths and areas) and the third classifies them by their mass (weight). With pipes (Class I) the classification is pipe material, nominal bore, and depth at which laid.

The entries in the divisions are termed 'descriptive features' as when three are linked together (one from each division), they normally provide a reasonably comprehensive description of a billed item. The CESMM3[1] classification does not, however, break down components into a large number of parts, which would result in excessively complicated coding arrangements. For example, it does not subdivide pipes according to pipe quality, type of joint, method of trenching or type of terrain. Nevertheless, the CESMM3[1] work classification provides the core of wider ranging descriptions with a consistent approach.[5]

Each Class contains a set of measurement, definition, coverage and additional description rules, which amplify and clarify the preceding information. It is vital to read these notes in conjunction with the preceding measurement particulars, since they may make reference to items that have to be included in the price of the measured item without the need for specific mention, or give matters that have to be inserted in the description of the measured item in the Bill.

Billed item descriptions are not required to follow precisely the wording in the work classification, although in many instances it will be advisable to do so. In measuring a joint in concrete made up of a rubber waterstop 175 mm wide, the billed description can state exactly that without the need to mention that the waterstop is 'plastics or rubber; width not exceeding 150–200 mm' as G652. On occasions an additional description rule accompanying the work classification will require more descriptive detail than is given in the classified lists, for instance the sizes and types of marker posts shall be stated in item descriptions (Class K; additional description rule A11).

The various rules override the tabulated classifications and it is better to simplify the wording of the work classification rather than to duplicate information, although the classification list will still provide the code number for the item where it is required. For example, the item description of a carriageway slab of DTp specified paving quality concrete, 150 mm deep, will not require the inclusion of concrete pavement from the first division of Class R, since it would be superfluous, nor would the addition of depth, 100–150 mm, be appropriate when the actual depth of the slab is required in additional description rule A1 (Class R), and to accord with the procedure described in 3.10.

Lists of up to eight different descriptive features are given in work classifications to cover the most commonly encountered items in the class, but they cannot encompass every conceivable alternative, hence subdivision 9 is left vacant to accommodate some other type or category.

The assembly of a description based on the three divisions of tabulated classification may not always be adequate, having regard to Section 5 of the CESMM3[1] and rules accompanying work classifications. It is helpful to separate the standard description from the additional information by a semicolon.

Billed descriptions of components for Permanent Works shall be concise and shall not include the processes of production or constructional techniques. An item of fabric reinforcement could read 'Mild steel fabric reinforcement type A252 to BS 4483, nominal mass: 3.95 kg/m^2', but certainly not as 'Supply, deliver, cut and fix mild steel fabric reinforcement type A252 to BS 4483, nominal mass: 3.95 kg/m^2'. However carefully such an item is drafted, there is a risk that an operation may be omitted, such as cleaning the reinforcement, and for which the Contractor might subsequently claim. It is more satisfactory to rely on the wording of the Specification, Drawings and Conditions of Contract together to provide the complete contractual requirements. Workmanship requirements should be written into the Specification and not the Bill of Quantities.

No billed item may contain more than one component from each division of a work classification list (3.4). Taking Earthworks (Class E) as an example, a single billed item cannot incorporate excavation for foundations, filling and disposal of excavated material, nor can it include both topsoil and rock in the same item, nor a combination of maximum depth ranges. A similar principle applies to the use of additional descriptions, which often stem from additional description rules accompanying the CESMM3[1] work classification. For instance in class I (Pipework — Pipes) additional description rule A2 requires materials, joint types, nominal bores and lining requirements of pipes to be stated in item descriptions. Hence variations in any of these components will result in separate billed items. This highlights the importance of rules in the work classifications—they are not merely explanatory comments and the rules of measurement, definition, coverage and additional description must be read very closely in conjunction with the tabulated classification lists. The measurement rules (3.6) identify any variation from the normal measurement approach described in rule 5.18; definition rules (3.7) clarify the meaning of terms used in the work classification; coverage rules (3.8) amplify the extent of the work to be priced in bill items; and the additional description rules (3.9 and 3.10) make provision for further descriptions and subdivisions of billed items. Some believe that the work classification is unduly inhibiting, but it does at least ensure a good measure of uniformity in bill descriptions.

The CESMM3[1] Work Classification is confined to Permanent Works and the rates inserted against these items in the Bill of Quantities will cover the costs that are proportional to the quantities of measured work. For instance, there is no item of measured work for bringing plant to and from the site. Where the cost of this activity is significant the Contractor could advantageously enter it as a method-related charge.[5]

The Work Classification prescribes the unit of measurement to be used for each billed item, ranging from stated sums and numbered items, linear items in metres, areas in square metres (hectares for general site clearance), volumes in cubic metres and weight (mass) in tonnes.

Coding and Numbering of Items

Section 4 of CESMM3[1] describes the coding system adopted in the Work Classification and how it can be used in the Bill of Quantities for numbering billed items. It is not, however, a requirement of CESMM3[1] that code numbers should be used as billed item numbers. Bills may be subdivided into parts covering different phases or sections of the work, such as the component parts of a sewage treatment works, and items with the same code number may be repeated in different parts of the bill. The different parts of the bill can be numbered and the part number can prefix the item number. Thus an item K152 in Part 6 of the bill would become 6.K152. Barnes[5] sees advantages to Contractors in the extensive use of CESMM3[1] coding as an aid to estimating and cost control and believes that it encourages uniformity of sequence of items in bills.

Each item in the Work Classification has been assigned a code number consisting of a letter and not more than three digits. The letter corresponds to the Work Class, such as E for Earthworks, and the digits relate to the relevant components in the first, second and third divisions of the class. An example will serve to illustrate its application.

Code H445 identifies an item as

class	H	precast concrete
first division	4	column
second division	4	length 10–15 m
third division	5	mass 2–5 t

In practice the description of this item will need amplifying to include the position in the Works, concrete specification, cross-section and principal dimensions, mark or type number and mass of the particular unit, as required by rules A1, A2, A4 and A6 of Class H.

The symbol * is used in the rules to the Work Classification to indicate all the numbers in the appropriate division, such as H 44 * representing the group of code numbers from H 441 to H 448 inclusive.

As a general rule billed items will be listed in order of ascending code number (4.3). The code numbers have no contractual significance (4.4). Some argue that the coded approach places too much emphasis on computer usage, with resultant constraints on bill compilation including an illogical sequence of items when related to normal quantity surveying practice.

Where a component of an item is not listed in the Work Classification, the digit 9 shall be used (4.5). The digit 0 can be used for divisions that are not applicable or where fewer than three divisions of classification are given (4.6). Suffix numbers can follow the code to cover varying additional descriptions of the type described earlier for precast concrete columns, when the code numbers would be H 445.1, H 445.2, H 445.3, etc.

Preparation of Bills of Quantities

The rules prescribed in Section 5 for the preparation of Bills of Quantities will also apply to the measurement of completed work (5.1).

Paragraph 5.2 prescribes a standard format for civil engineering bills of quantities to ensure uniformity of presentation. For example, a bill of quantities for a riverworks contract for a new power station might contain the following.

Section A	List of principal quantities
Section B	Preamble
Section C	Daywork schedule
Section D	Work items
	Part 1 General items
	Part 2 Demolition and siteworks
	Part 3 Access roads
	Part 4 Pipework
	Part 5 Dredging
	Part 6 Pump chambers
	Part 7 Circulating water ducts
	Part 8 Wharf wall
	Part 9 Jetty
Section E	Grand summary

The sections are identified by letters to distinguish them from the locational or cost-significant parts of the Works, which have reference numbers.

List of Principal Quantities

Paragraph 5.3 advocates the inclusion of a list of principal quantities, being the main components of the Works, with their approximate estimated quantities, so that tenderers obtain an overall picture of the general scale and character of the proposed Works at the outset. It is expressly given *solely* for this purpose with the intention of avoiding any possible claims on account of divergences between the list of principal quantities, or the impression given thereby, and the detailed contents of the Bill of Quantities. Nevertheless, some quantity surveyors feel that the inclusion of such a list on a remeasure type of contract may be of limited value and could be contentious. However, this list will assist the contractor in determining whether he has the resources to carry out the work.

The list can be kept relatively brief and should not usually exceed one page in length. It is best prepared from the draft bill, although it is not essential to subdivide the list into the bill parts. The amount of detail given will vary with the type and size of contract.

A list of principal quantities relating to a reservoir follows.

Part		
1	General Items	
	Provisional Sums	£20 000
	PC Items	£76 000
2	Reservoir	
	excavation	15 000 m³
	concrete	12 500 m³
	formwork	6 600 m²
	steel reinforcement	150 t
3	Pipework	
	pipelines	850 m
	valves	22 nr
4	Embankment	
	filling	12 000 m³
5	Access Road	
	concrete road slab	1 800 m²
6	Fencing	
	mild steel fence	1 400 m

Preamble

The preamble in a civil engineering bill of quantities is to indicate to tendering contractors whether methods of measurement other than CESMM3[1] have been used for any part of the Works and whether any modifications have been made in applying CESMM3[1] to meet special needs where there are important practical reasons for adopting a different procedure (5.4). Circumstances in which a different procedure may be used include the introduction of performance specifications resulting in less detailed measurement, and the use of permitted alternatives, as sometimes adopted for highway and tunnelling contracts, and contractor-designed work, and possibly involving the use of non-standard rules for measurement.[5] Wherever practicable the unamended CESMM3[1] should be used in the interests of uniformity.

The majority of civil engineering contracts include work below ground in general excavation, trenching for pipes, boring or driving, and in other ways. Paragraph 5.5 requires a definition of rock to be included in the Preamble in these circumstances, and it is usually related to geological strata as in the DTp *Specification for Highway Works*.[31] Alternative or complementary approaches include prescribing a minimum size of boulder (0.20 m³ in the DTp Specification and 1 m³ in rule M8 of Class E of CESMM3,[1] except that the minimum volume shall be 0.25 m³ where the net width of excavation is less than 2 m), and strata which necessitate the use of blasting or approved pneumatic tools for their removal. This definition can be of considerable significance in determining whether or not additional payments shall be made for excavation, boring or driving work.

Daywork Schedule

It is necessary to make provision for a daywork evaluation of work which cannot be assessed at bill rates or rates analogous thereto. CESMM3[1] provides three alternative procedures (5.6 and 5.7).

(1) A list of the various classes of labour, material and plant for which daywork rates or prices are to be inserted by the tenderer.
(2) Provision for payment at the rates and prices and percentage additions contained in the current Federation of Civil Engineering Contractors Schedules of Dayworks,[27] adjusted by the Contractor's percentage additions or deductions for labour, materials, plant and supplementary charges.
(3) The insertion of provisional sums in Class A of the bill of quantities for work executed on a daywork basis comprising separate items for labour, materials, plant and supplementary charges, and applying the appropriate percentage addition or deduction, as prescribed in the second method, to each provisional sum.

The third method is felt by the author to offer the most advantages, since it directly influences the Tender Total, thus maintaining an element of competition, at the same time providing a widely known and accepted basis of computation which is easily implemented.

Work Items

Division into Parts The Bill of Quantites is divided into sections in accordance with paragraph 5.2 and Section D contains work items which may be arranged into numbered parts, and which will differ from one bill to another. The division into parts is mainly determined by the main components of the project, locational considerations, limitations on the timing or sequence of the work, and it enables the person preparing the bill to distinguish between parts of the work which are thought likely to give rise to different methods of construction or considerations of cost (5.8). This form of division will extend the usefulness of the Bill for estimating purposes and in the subsequent financial control of the contract. Sound division into parts requires knowledge of the factors influencing the Contractor's costs and assists in promoting more positive working relationships between the Engineer and the Contractor.[5]

General items (Class A) may be grouped as a separate part of the bill.

Items in each part shall be arranged in the general order of the Work
Classification (5.8), even though this sequence is rather illogical in
some instances. It will be noted that the bill *may* be arranged into
numbered parts and that it is not therefore obligatory.

Headings and Sub-headings

Paragraph 5.9 prescribes that each part of the bill shall be given a
heading and that each part may be further subdivided by sub-
headings, all inserted as part of item descriptions. For the sake of
clarity, a line shall be drawn across the item description column below
the last item to which the heading or sub-heading applies, and headings and sub-headings shall be repeated at the start of each new page
listing appropriate items (5.9).

Itemisation and Description

All work shall be itemised in the bill with the descriptions framed in
accordance with the Work Classification. However, paragraph 5.10
states that item descriptions may be extended or work subdivided into
a larger number of separate items than required by CESMM3[1] if it is
thought likely that the work will give rise to special methods of construction or considerations of cost. The word 'may' is used to prevent
the Contractor having a basis of claim for extra payment if the previous
assumptions proved incorrect, and to permit flexibility of approach in
the preparation of bills.

The Work Classification should not be applied too rigidly: it provides standardised minimum information and essential guidelines for
the benefit of contractors, but permits the inclusion of more information where deemed advisable or desirable, particularly where cost
significant aspects are involved.[5]

Descriptions

Item descriptions only *identify* work whose nature and extent is defined by the contract documents as a whole, including the Drawings,
Specification and Conditions of Contract (5.11). Unlike Bills of Quantities for building work, they do not contain all the information needed
for pricing the Work, and continual reference to the Drawings and
Specification is essential. This policy encourages the complete design
of work prior to inviting tenders, and the cost of civil engineering work
frequently depends extensively on the shape, complexity and location
of the work and the nature of the terrain; this information is best
extracted from the Drawings. To assist in the process of identification

CESMM3[1] often requires the inclusion of information on location or other physical features (5.13). Paragraph 5.12 extends this approach by stating that any descriptive information required by the Work Classification may be replaced by reference to the appropriate Drawings or Specification clauses.

Barnes[5] has suggested that CESMM3[1] classes and codes could be used as numbering or referencing systems for Specifications to assist in cross-referencing. He also advocates the use of drawing numbers in bill items or bill sub-headings.

The use of measurement procedures at variance with those prescribed in CESMM3[1] without express exclusions might possibly be construed as 'errors' and treated as variations in accordance with clause 55(2) of the ICE Conditions of Contract.[13] Hence it is important that the requirements of CESMM3[1] should be closely followed in drafting civil engineering bills of quantities. Situations will arise where no guidance is given by CESMM3[1] and a suitable non-standard item should be drafted.

The cost of similar work in different locations can vary considerably, and in these circumstances locational details should be inserted in the bill items to permit the Contractor to adjust his rates accordingly. For instance, reinforced concrete of the same mix to be laid in the base of a pump sump, in a floor slab at ground level, or in the tank base to a water tower, are all similar forms of construction but carried out under entirely different conditions, resulting in considerably different costs and creating the need for additional descriptive information in the bill items.

Ranges of Dimensions

Where the Work Classification prescribes a range of dimensions for a component, but the component in question is of one dimension, this dimension shall be stated in the item description in place of the prescribed range (5.14). For instance, the placing of *in situ* concrete suspended slabs in Class F have four ranges of thickness listed in the third division. When measuring slabs all having a thickness of 200 mm, the thickness given in the item description would be 200 mm in place of the CESMM3[1] range of 150–300 mm (exceeding 150 mm and not exceeding 300 mm).

Prime Cost Items

Prime Cost (PC) Items shall be inserted in a Bill of Quantities under General Items (Class A) to cover work carried out by Nominated Subcontractors and each prime cost is to be followed by two further items.

(1) Labours in connection therewith including site services provided by the main Contractor for his own use and which he also makes available for use by the Nominated Sub-contractor and which are listed in the CESMM3[1] (5.15), such as temporary roads, scaffolding, hoists, messrooms, sanitary accommodation and welfare facilities, working space, disposal of rubbish and provision of light and water. Where the Nominated Sub-contractor is not to carry out work on the site, the item shall include unloading, storing and hoisting materials supplied by him and the return of packing materials (5.15).
(2) A further item expressed as a percentage of the PC Item to cover all other charges and profit (5.15).

This procedure follows closely the arrangements detailed in the ICE Conditions of Contract[13] (clauses 58 and 59). Any special labours beyond those specified in 5.15(a) must be included in item descriptions. Where substantial special attendance facilities are envisaged but cannot be assessed precisely, these should be incorporated in Provisional Sums. Barnes[5] has described how the fixing of materials for work of Nominated Sub-contractors by the main Contractor is not covered by Prime Cost Items nor by attendance, other charges and profit items.

Provisional Sums

Provisional Sums shall be used to cover contingencies of various types and they can be entered in various sections of the Bill of Quantities for subsequent adjustment. Items for specific contingencies are to be included in the General Items (Class A — 4.2), while other items may be included in other classes to cover, for instance, possible extensions of work (5.17). A general contingency sum shall be entered in the Grand Summary in accordance with paragraph 5.25.

The omission of provisional quantities means that the quantities in the Bill are to be measured accurately from the Drawings and are to be the best possible forecast of the nature and extent of the work which will actually be required, including such items as excavation of soft spots and hours of pumping plant.[5]

The ICE Conditions of Contract[13] [56(2)] provides for bill rates to be increased or decreased if they are rendered unreasonable or inapplicable as a result of the actual quantities being greater or less than those stated in the bill. It will be appreciated that the cost of one billed item frequently depends on its relationship to others, and the cost of the whole work can change substantially if the relative proportions of the quantities are varied.[5]

The inclusion of Provisional Sums to cover possible additional work will result in appropriate rates for the work subsequently having to be negotiated.

Quantities

Paragraph 5.18 prescribes that 'quantities shall be computed net from the Drawings'. The billed quantities will be the lengths, areas, volumes and masses of the finished work which the Contractor is required to produce, with no allowance for bulking, shrinkage or waste. The only exceptions are where CESMM3[1] or the Contract contain conventions for measurement in special cases, such as the volume of concrete is to include that occupied by reinforcement and other metal sections (rule M1 of Class F).

In the original *Standard Method of Measurement of Civil Engineering Quantities* some items were measured 'extra over' others — that is, the price for the second item covered only additional costs over the first. For example, facings and fair-faced brickwork were measured as extra over the cost of ordinary brickwork, thus eliminating the need to deduct the ordinary brickwork, displaced by faced work. Similarly bends, junctions and other fittings were measured extra over pipe sewers and drains. The three editions of the CESMM discontinued this practice. Brickwork faced on one face will be measured as a composite item, while pipe lengths are measured along their centre lines and shall include lengths occupied by fittings and valves (rule M3 of Class I), but the fittings and valves are not measured extra over, although the estimator will need to deduct the costs of the lengths of displaced pipe when estimating the rates for the enumerated fittings and valves (Class J).

Quantities are usually rounded up or down (half a unit or more up and less than half a unit down) to avoid the use of fractional quantities in the bill. Where fractional quantities are used, because of the high unit cost of the item, they should be restricted to one place of decimals (5.18).

Units of Measurement

Paragraph 5.19 lists the units of measurement used in civil engineering bills of quantities with their standard abbreviations. Square metres are abbreviated to m^2 or m2 and not sq m, and cubic metres to m^3 or m3 and not cu m. Number is represented by nr and not no. The term 'sum' is used where there is no quantity entered against an item. The abbreviations are restricted to not more than three characters, contain no capital initials and are not followed by a full stop.

Work affected by Water

Paragraph 5.20 prescribes that where an existing body of open water (other than groundwater) such as a river, stream, canal, lake or body of tidal water is either on the site or bounds the site, each body of water shall be identified in the preamble to the bill of quantities. A reference shall also be given to a drawing indicating the boundaries and surface level of each body of water or, where the boundaries and surface levels fluctuate, their anticipated ranges of fluctuation.

A typical preamble clause could be 'The Site is bounded by the Beeston Canal whose location is shown on Drawing BSDW 12B. The width of the canal is constant and it is anticipated that the surface level may fluctuate between 85.30 and 86.00 AOD.' Excavation work below water requires a suitable reference to the preamble clause in the item description (rules A2 and M7 of Class E).

Ground and Excavation Levels

With excavation, boring or driving work, it is necessary to define the 'Commencing Surface' where it is not also the 'Original Surface', and the 'Excavated Surface' where it is not also the 'Final Surface' (5.21). In most cases, however, these intermediate surfaces will not be mentioned in the descriptions, and it will then be assumed that the item covers the full depth from the 'Original Surface' (before any work in the Contract is commenced) to the 'Final Surface' (when all work shown on the drawings has been executed).[3]

The definitions of these terms are given in paragraphs 1.10–1.13 inclusive. The surfaces do not have to be identified by a level, as for instance above Ordnance Survey Datum; provided the descriptions are clear and practical they are satisfactory. For instance, such descriptions as '300 mm below Original Surface' and '150 mm above forma-

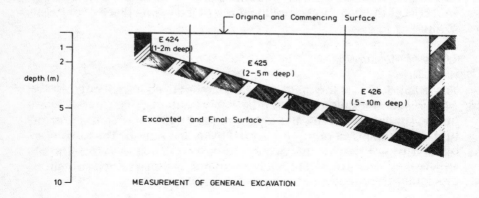

MEASUREMENT OF GENERAL EXCAVATION

tion' are acceptable. Excavation and similar work is classified according to a range embracing its total depth, as shown in the previous sketch.

Form and Setting of Bills of Quantities

The Bill of Quantities should desirably have the format on A4 size paper described in paragraph 5.22 and as illustrated below (the quantity, rate and amount: £ columns shall each have a capacity of ten million less one, and a binding margin should be provided).

Number	Item Description	Unit	Quantity	Rate	Amount	
					£	p
20	90	10	20	20	20	8
widths (in mm)						
				Page total		

The amounts on each page shall be totalled and these totals carried to a summary at the end of each part of the bill. The total of each part is transferred to the Grand Summary (5.23 and 5.24). The Part Summaries are normally followed by the *General Contingency Allowance*, which is a Provisional Sum inserted to cover unforeseen work (5.25), an *Adjustment Item* (5.26) and finally the Grand Total (total of the priced bill of quantities) [5.27].

Completion, Pricing and Use of the Bill of Quantities

Paragraph 6.1 prescribes that rates and prices shall be inserted in the rate column in pounds sterling with pence entered as decimal fractions of one pound. The generally accepted practice is to regard figures in the 'rate' column as rates and those in the 'amount' column as prices on the supposition that quantity × rate = price. The prices in each part are totalled and transferred to the Grand Summary in the manner previously described.

An *Adjustment Item* is included in the Grand Summary as a convenient place where the tenderer can make an adjustment without having to alter rates or amounts against work items. The final tender adjudication is normally undertaken by senior estimating and management staff, who are particularly concerned with assessment of the risk involved and pricing levels. Previously such adjustments were often made to some of the larger prices entered against preliminary or general items in the Bill. The inclusion of the Adjustment Item to incorporate a lump sum addition or deduction as the last item in the Grand Summary is a much more satisfactory arrangement for both Contractor and Employer (6.3).

The Adjustment Item is a fixed lump sum and the sum entered against this item is not adjustable for variations in the Contract Sum, although it is subject to adjustment when the Baxter adjustment formula is incorporated to deal with labour and material price fluctuations (6.5). The Baxter formula superimposes price fluctuation resulting from index movement over adjustments in the Contract Sum assessed under the ordinary Contract Conditions.[5]

Payment of the Adjustment Item shall be made by instalments in interim certificates in the proportion that the value of certified permanent and temporary works [sub-clause 60(2)(a) of the ICE Conditions of Contract[13]] bears to the total of the Bill of Quantities, before the addition or deduction of the Adjustment Item, and with no retention money deducted, and a statement to this effect shall appear in the Preamble. These payments shall not exceed in aggregate the amount of the Adjustment Item and any final balance due is certified in the next certificate prepared after the issue of the Certificate of Substantial Completion for the whole of the Works under clause 48 of the Conditions of Contract (6.4).

4 Method-related Charges and Pricing and Measurement of Civil Engineering Work

Method-related Charges (CESMM3 Section 7)

Underlying Philosophy

The valuation of variations and agreement to the cost of delays have generally been the main causes of dispute on civil engineering contracts. Contractors frequently claim that the measured quantities of permanent work priced at billed rates do not represent the true value of constructional work where significant variations have occurred. Engineers, on the other hand, generally believe that the priced bill of quantities represents a shopping list of items, and on completion of construction the work can be remeasured and valued at the billed rates. That the latter is unsatisfactory is evidenced by the large claims settlements agreed on many projects, where variations and unforeseen physical conditions or artificial obstructions occur.

Many of the costs arising from civil engineering operations are not proportional to the quantity of the resulting permanent work. It cannot really be a sound approach to recover the cost of bringing a tower crane on to the site and its hire, operation and subsequent removal by hidden costs in the 'preliminaries', where provided, or in the cost per cubic metre of the various work sections for which the crane was used. It is believed by the drafting committee that method-related charges provide a better way of representing the Contractor's site operation costs, such as the provision of site accommodation and temporary works, and the setting up of labour gangs, sometimes described as 'site mobilisation'.

Objectives

Certain clauses in the original *Standard Method of Measurement of Civil Engineering Quantities*, such as site investigation, post-

59

tensioning and *in situ* piling provided items for plant provision and removal. These caused inconsistency, since many other categories of permanent work included no such provision. CESMM3[1] helps to remove these inconsistencies by the use of method-related charges. The Contractor can enter and price such costs as he considers he cannot recover through measured rates, such as site accommodation, site services, plant, temporary works, supervision and labour items — all at the tenderer's discretion.

Accepting that expertise in design rests with the Engineer, it seems equally evident that expertise in construction methods lies with the Contractor. It is accordingly logical that the Contractor should be able to decide the method of carrying out the works. A blank section in the Bill of Quantities will permit him to list, describe and price these items.

CESMM3[1] does not make the use of method-related charges compulsory but the sponsors see great merit in their use through easier evaluation of variations, a more stable and realistic cash flow to the Contractor and by directing the Engineer's attention to the basis of construction costs, to lead to more rational designs. In the event of an Employer using CESMM3[1] as the basis for measurement but excluding method-related charges, he should insert a note in the bill preamble stating that Section 7 of CESMM3 shall not apply.[5]

The extent of temporary works on a civil engineering contract is often enormous and to spread their costs over unit rates must be unsatisfactory, since so few of them are proportional to the quantities of permanent work. Hence it is believed by the sponsors to be beneficial to all parties for the Contractor to have the opportunity to insert these costs, properly itemised in a separate part of the bill.

Division into Time-related and Fixed Charges

There are two basic types of method-related charge: time-related charges and fixed charges (7.1). The cost of bringing an item of plant on to a site and its subsequent removal is a fixed charge and its running cost is a time-related charge. The tenderer is requested to distinguish between these charges and must fully describe them so that the coverage of the items entered is positively identified. The fixed costs are not related to quantity or time. With time-related items, the Contractor is to enter full descriptions of the items and their cost, but not the timing or duration, as they will not be charged at weekly or monthly rates.[5] The Quantity Surveyor or Engineer preparing the bill will provide adequate space and the Contractor will enter the particulars in the description column, preferably using the order of classification and descriptions adopted in Class A (7.3). Typical Bill entries covering both types of method-related charge are given on pages 67–69.

Application of Method-related Charges

Problems may arise in practice where Contractors entering method-related charges in a bill fail to identify them as fixed or time-related, or fail to describe them adequately. This will involve the Quantity Surveyor or Engineer in constant checking of entries.

The items entered as method-related charges are not subject to admeasurement, although the Contractor will be paid for these charges in interim valuations in the same way as he is paid for measured work, and a statement to this effect shall appear in the Preamble (7.6 and 7.7). Hence, in the absence of variations ordered by the Engineer, the sums entered against method-related charges will re-appear in the final account, and will not be changed merely as a result of the quantity of method-related work carried out being different from that originally estimated by the tenderer. The valuation of these charges should be made easier by their division into time-related and fixed costs. The Contractor will not be obliged to construct the works using the methods or techniques listed in his method-related charges, but he will nevertheless be paid as though the techniques indicated had been adopted (7.8). For example, if the Contractor inserted charges for a concrete batching plant and subsequently used ready-mixed concrete, the appropriate interim payments will be distributed over the quantity of concrete placed. If, however, changes in techniques are instructed by the Engineer, then these changes will be paid for as variations. Method-related charges shall, however, like the Adjustment Item, be subject to Baxter formula price adjustment.[5]

The introduction of method-related charges thus enables the Contractor at his option to enter separately in his tender such non-quantity proportional charges as he considers will have a significant influence on the cost of the work, and for which he has not allowed in the rates and prices for other items (7.2). Where the Contractor omits to enter any method-related charges and merely prices the contract as a traditional bill of quantities, these charges will be deemed.to have been included in the pricing of other items and the tender will not be invalidated. If the Contractor enters a method-related charge for an item that cannot be performed or if the cost is obviously incorrect, the Engineer should draw the attention of the Contractor to the mistake and give him the opportunity to withdraw his tender.

All method-related charges must be fully described (7.4), including the resources expected to be used and the particular items of Permanent or Temporary Works to which the item relates, although a method can be changed subsequently (7.5). It is in the Contractor's interest to be explicit in his descriptions to secure prompt payment. Barnes[5] postulates that where activities on the site resemble those described in the method-related charges, payment can be made promptly, otherwise the Engineer may have grounds for withholding payment until the related permanent work is complete.

Alongside the description of each method-related charge the tenderer will enter whether it is fixed or time-related. In interim valuations, payment of fixed charge items will be made when the operation so described, such as erection of site offices, has been completed. Time-related charges for items such as maintaining a temporary access road will be paid monthly, according to the Engineer's assessment of the proportion of total time that has elapsed at the date of assessment. If the Contractor enters the operation of a tower crane for 22 weeks as a time-related charge and then proceeds to use it on the site for 28 weeks, no additional payment or adjustment will necessarily follow. Conversely, as described by Barnes,[5] if a method-related charge covered de-watering plant be subsequently rendered unnecessary by a drought, the Contractor is still eligible for the payment because he bore the risk of having to do an indeterminate amount of work, the payment being *pro rata* to the proportion of associated Permanent Work completed. Only in the case of variations wll the price be subject to adjustment. If a variation increases work volume or causes delay, and thereby results in an increase in the cost of method-related charges, these can be adjusted within the terms of the Contract, without the necessity of a claim. On occasions a variation deleting work could result in a reduction in a charge. Much depends on whether the variation actually extends or shortens the time for which the resources are required.

Class A shows specified requirements that the Engineer or Quantity Surveyor may insert in the bill, followed by a list of items which the Contractor may enter, but is not compelled to do so. He will choose to do so if he considers that it provides a more realistic basis for pricing. The danger is that he could enter almost any items except materials and so abuse the arrangements. In practice, site accommodation, services and temporary works have been the items most commonly inserted and plant to a lesser extent. Multi-purpose types of plant like tower cranes, derrick systems and ropeways are very suitable items, since they are used for hoisting a variety of materials and involve both fixed and time-related charges.

With the payment of time-related charges the Engineer is not tied to the dates shown on the programme and will have regard to the date when the items were actually provided on the site. Interim payments for time-related charges should be proportional to the extent of satisfactory completion of the particular activity, and the Engineer will need to assess the total period over which the charge should be spread. The Contractor should ensure that the descriptions and durations of time-

related charges match the information given in the contract programme, thus assisting in substantiating the Engineer's assessment of a
reasonable proportion for payment. In the event of variations, time-
related charges require adjustment if rendered unreasonable or inapplicable; the more precise the descriptions, the more realistic the
adjustments.[5]

Advantages and Disadvantages of Method-related Charges Approach

The use of method-related charges should remove substantial sums of
construction costs, which do not vary in proportion to the volume of
permanent works executed, from the pricing of these permanent
works and so reduce likely claims. If used effectively they should
enable the Contractor to recover in monthly valuations the cost of
items other than permanent work on an equitable basis, either in the
event of work proceeding largely as planned at tender stage or in the
event of substantial variations.

In either case the Contractor will be able to recover these non-
quantity proportional items on a monthly basis and not be obliged to
wait until the end of the contract, or at least the later stages, to submit
claims to recover costs.

The Employer should have the benefit of a more accurate valuation
of variations with improved monitoring of the financial position of the
contract. He will also be aware of the level of expenditure at an earlier
stage, which will help him to plan his cash flow and budget for his
ultimate level of financial commitments. It is interesting to note that in
the majority of contracts, incorporating method-related charges operating during the currency of the earlier editions of the CESMM, their
use proved beneficial.

It is claimed that, although there can still be different approaches to
the pricing of bills, by introducing method-related charges the different bases for computation and approach are more clearly identified.
The Employer thus gains by the use of a cost structure that is better
suited to deal with variations and changes, while the Contractor receives more prompt and equitable payment. No significant problems
have occurred with final account preparation, and this approach assists
the Contractor in making claims and the Engineer in settling them on a
more realistic basis with less argument and conflict. When method-
related charges are not used, differing policies for pricing and allocat-
ing indirect costs are hidden within the wide variation in measured
work unit rates.

The success of the method-related charges approach is largely dependent on its sensible use by both Contractor and Engineer. If
Contractors refuse to enter method-related charges and Quantity

Surveyors and Engineers are slow in authorising payment of them under sub-clause 60(i)(d) of the Conditions of Contract,[13] then little will be achieved, particularly in easing the Contractor's cash flow problems.

The principal danger is generally thought to be one of high early payments with no collateral security, although it must be accepted that the chance of a Contractor becoming bankrupt at the start of a project is very small and the Employer is not without his safeguards. Although Barnes[5] argues that loaded method-related charges are more easily identifiable than loaded unit rates, others fear that method-related charges could be used by Contractors to qualify tenders and that deliberate mistakes could be included to give the tenderer a second chance after receipt of tenders. Furthermore, Engineers might experience some difficulty in coping with large numbers of awkwardly worded method-related charges. One suggested solution is for Contractors to be asked to submit lists of unpriced method-related items for approval some time before the tender date. This would, however, probably lead to a lengthening of tendering periods and could cause administrative problems.

On the whole the use of method-related charges, although not without their dangers, can lead to improved design, estimating, tender selection, contract administration and cost control techniques, with better cash flow to the Contractor, particularly in the early part of the Contract.

Class A: General Items

General Items contained in Class A of the CESMM3[1] are items, other than Permanent Works, which the Contractor is required or chooses to do or provide. There are some exceptions to this general rule, such as Provisional Sums and Prime Cost Items related to Permanent Works, and the non-inclusion of some Temporary Works, such as formwork and temporary support of tunnel work. In the past many of these items were often termed 'Preliminaries'.

General Items shall each be described as quantity-related, time-related or fixed, to simplify arrangements for subsequent certification and payment. This part of the bill contains items covering the Contractor's obligations under the Contract and all the services that he will be required to provide.

The insurance items are listed in A 12–3, including insurance of the Works and constructional plant, and third party insurance. These constitute the contractual requirements under clauses 21 and 23 of the ICE Conditions of Contract.[13]

A2 covers requirements specified by the Engineer which the Contractor is bound to meet, whereas the method-related charges entered in A3 are items which the tendering Contractor chooses to insert. Rule D1 of Class A defines 'specified requirements' as all work, other than the Permanent Works, which is expressly stated in the Contract (Specification or other contract document) to be carried out by the Contractor and of which the nature and extent is defined. This ensures that the Contractor's attention is drawn to contractual requirements which in other circumstances might be a matter for his own decision. For example, pumping and de-watering appear in A 276 and A 277 respectively as specified requirements to be inserted by the Engineer, and are also listed in A 356 and A 357 as method-related charges to be inserted by the Contractor, depending on the specific requirements of the particular contract.

Another function of specified requirements items is to obtain a price which can be adjusted in the event of a variation. Otherwise, if for instance the Engineer is to change the details of accommodation for the Engineer's Staff, in the absence of a specific item, adjustment of an undisclosed price could prove difficult.

Rule M2 of Class A requires that a quantity shall be inserted against all specified requirements items for which the value is to be determined by admeasurement. While rule A2 of Class A requires the establishment and removal of services or facilities to be distinguished from their continuing operation or maintenance. Items involving fixed costs will usually be sums that are not subject to quantified measurement and are often described as 'establishment and removal of'. By comparison time-related items may either be quantified such as in hours or weeks, or given as sums. These alternative approaches are illustrated in the examples that follow. A unit of quantity should be inserted when the cost is directly proportional to a measurable quantity such as the number of sets of progress photographs, but the insertion of a sum for maintenance of offices for the Engineer's staff.

Testing items should include particulars of samples and methods of testing, although some testing may be listed separately where prescribed under other classes. Tests involving the assembly or construction of substantial testing facilities require an 'establishment of testing facilities' item.[5]

The Method-related charges division of Class A lists a number of the more common items, but these do not restrict the Contractor in any way. The Contractor can insert, in the space provided in the General Items part of the Bill, other items which will not be proportional to the quantities of the Permanent Works, distinguishing between time-related and fixed charges. Some typical entries are illustrated on page 69.

The time element of time-related charges can be expressed in various ways, but such expressions as 'the duration of construction' or 'completion of the wharf' are generally more useful than stated lengths of time which are not related to site activities and which may prove more difficult to apply to changed conditions.

Provisional Sums can be incorporated in accordance with paragraph 5.7 (items A 411–A 416), followed by any other relevant Provisional Sums as given on page 70. Prime Cost Items are grouped under A5 and A6 in the General Items. Each Prime Cost Item shall be followed by an item for labours in connection therewith and an item for other charges and profit in accordance with paragraph 5.15. Where special labours are required, these shall be described adequately including the period of time involved.

Some typical General Items follow, referenced in accordance with CESMM3[1] (Class A).

Number	Item Description	Unit	Quantity	Rate	Amount £	p
	PART I GENERAL ITEMS					
	Contractual requirements					
A110	Performance bond.	sum				
A120	Insurance of the Works.	sum				
A130	Third party insurance.	sum				
	Specified requirements					
	Accommodation for					
	Engineer's staff					
A211.1	Establishment and removal of office for the Engineer's staff, as Specification clause A25.	sum				
A211.2	Maintenance of offices for the Engineer's staff.	sum				
A211.3	Maintenance of offices for the Engineer's staff after the issue of the Completion Certificate.	wk	15			
	Attendance upon Engineer's staff					
A242	Attendance upon the Engineer's staff; chainmen.	wk	80			
	Testing of materials					
A250	Testing of materials; concrete test cubes as Specification clause C86.	nr	300			
	1/1		Page total			

Civil Engineering Quantities

General Items

Number	Item Description	Unit	Quantity	Rate	£	p
					Amount	
	Specified requirements					
	Testing of the Works					
A260.1	Clay pipes, nominal bore 150 mm, length 1380 m, as Specification clause I21.1.	sum				
A260.2	Clay pipes, nominal bore 225 mm, length 2540 m, as Specification clause I21.2.	sum				
A260.3	Clay pipes, nominal bore 300 mm, length 1170 m, as Specification clause I21.3.	sum				
	Temporary Works					
A272.1	Traffic regulation: establishment and removal as Specification clause A37.	sum				
A272.2	Traffic regulation: continuing operation and maintenance as Specification clause A37.	wk	80			
A276.1	Pumping plant: establishment and removal as Specification clause A52.	sum				
A276.2	Pumping plant: operation and maintenance as Specification clause A52.	h	600			
	1/2			Page total		

General Items
(Blank pages in Bill of Quantities — entries of method-related
charges made by tendering Contractor; *normally handwritten*)

Number	Item Description	Unit	Quantity	Rate	Amount £	p
	Method-related charges *Accommodation and buildings*					
A311.1	Establish offices: fixed.	sum				
A311.2	Maintain offices for duration of construction: time-related.	sum				
A311.3	Remove offices: fixed.	sum				
	Services					
A321.1	Establish electricity supply and standby generator: fixed.	sum				
A321.2	Provision of electricity for duration of construction: time-related.	sum				
	Plant *30 t crane for excavation and concreting of settling tanks*					
A.331.1	Bring to site: fixed.	sum				
A.331.2	Operate and maintain: time-related.	sum				
A.331.3	Remove: fixed.	sum				
	Temporary works *Compressed air for tunnelling from access shaft 7 to access shaft 10*					
A358.1	Establish compressed air plant: fixed.	sum				
A358.2	Compressed air supply: time-related.	sum				
A358.3	Remove compressed air plant: fixed.	sum				
	Supervision and labour					
A371	Management and supervision for duration of construction: time-related.	sum				
A373	Labour for maintenance of plant and site services for duration of construction: time-related.	sum				
1/3				Page total		

General Items — Provisional Sums and Prime Cost Items

Number	Item Description	Unit	Quantity	Rate	Amount £	p
	Provisional Sums					
	Daywork					
A411	Labour.	sum			60 000	00
A412	Percentage adjustment to Provisional Sum for Daywork labour.	%				
A413	Materials.	sum			30 000	00
A414	Percentage adjustment to Provisional Sum for Daywork materials.	%				
A415	Plant.	sum			30 000	00
A416	Percentage adjustment to Provisional Sum for Daywork plant.	%				
A417	Supplementary charges.	sum			25 000	00
A418	Percentage adjustment to Provisional Sum for Daywork supplementary charges.	%				
	Other Provisional Sums					
A420.1	Permanent diversion or support of existing services.	sum			18 000	00
A420.2	Repairs to existing structures and plant.	sum			25 000	00
	Nominated Sub-contracts which include work on the Site					
A510.1	Pumping plant.	sum			30 000	00
A520.1	Labours.	sum				
A540.1	Other charges and profit.	%				
A510.2	Lighting installation.	sum			13 000	00
A520.2	Labours.	sum				
A540.2	Other charges and profit.	%				
	Nominated Sub-contracts which do not include work on the Site					
A610.1	Bollards to wharf (20 nr).	sum			10 000	00
A620.1	Labours.	sum				
A640.1	Other charges and profit.	%				
A610.2	Rubber buffers to fender piles (200 nr).	sum			5 000	00
A620.2	Labours.	sum				
A640.2	Other charges and profit.	%				
	1/4			Page total		

Effect of CESMM3 on Pricing

The CESMM3[1] permits greater standardisation in the format of bills of quantities and this assists Contractors in pricing. The drafting committee believe that bills prepared under CESMM3[1] are more consistent, with work adequately itemised and described to include cost-significant items with a consistent level of detail. It is also stated that the coding system permits estimating, valuation, purchasing and cost control to use the same numerical references and that these will also simplify computerised data processing. These advantages are clearly identifiable in the operation of CESMM2.

General Items

Where the general items have been satisfactorily listed and priced and full use made by the Contractor of method-related charges, then the pricing of measured work is simpler. In theory all services and facilities that are not proportional to the quantities of permanent work will have been separated for pricing purposes. In practice all sorts of permutations are possible.

One Contractor may not insert any method-related charge items, while another could include for each and every type of plant, as he is in fact encouraged to do by Class A3, which, for instance, lists earth compaction plant and concrete mixing plant. Both of these types of plant involve some fixed costs in transporting plant and off-loading together with the subsequent transporting to a depot or another site. In between these fixed events the plant is operated and maintained and these can be regarded as time-related charges. On the other hand they could equally justifiably be considered as essential components of the earth compaction or concrete production, since they are proportional to the quantity of permanent work and hence could be more realistically included in the measured work rate.

These possible variations in approach in dealing with the pricing of plant could result in considerable differences in measured rates and general items. Hence the analysis, checking and comparison of priced bills by Quantity Surveyors and Engineers may not always be simplified significantly.

Measured Work

It took time for estimators to become familiar with the earlier editions of the CESMM and all its ramifications, particularly the rather confusing bill descriptions that flow from the standardised approach. There was a danger, particularly in the early period of its use that estimators might omit to include for items of work that are included in the particular bill item without the need for specific mention. Excavation and backfill to pipe trenches included in the pipework provide a specific example, even although beds, haunches and surrounds are measured separately (Class L).

A typical earthwork bill item could read as follows.

E424 General excavation in natural material to a maximum depth not exceeding 1–2 m

The rate would have to cover the following items additional to cubic excavation, which are deemed to be included by rules in Class E.

(1) Additional excavation needed for working space and removal of dead services.
(2) Upholding sides of excavation.

The estimator has therefore to proceed with caution and there is a danger that estimators may feel obliged to make some allowance in their tenders for the additional risk involved.

Where there is a wide range of different manhole types, sizes and connections, the enumerated approach to the measurement of manholes is not sufficiently sensitive. Similar criticism could be levelled at the maximum trench depth ranges used for excavation pricing purposes.

The *formwork* to concrete items (Class G) is separated according to a range of widths from not exceeding 0.1 m (100 mm) to exceeding 1.22 m (determined by width of plywood sheets). The underlying philosophy is that the separation of formwork in this way will help to indicate the relative complexity of the work and assist in the valuation of variations. The approach to the measurement of *in situ concrete* (Class F) separating provision from placing is eminently sensible. With the placing due attention can be paid to the specific location, which is important in pricing concrete in a widely differing set of situations, each generating different costs.

With *Piling* (Class P), three items of information are given for groups of cast in place concrete piles, preformed concrete and timber and isolated steel piles to bring out the cost significant items in this class of work.

(1) Cost related to number of piles (moving the rig and setting up piles).
(2) Length of pile (cost of materials).
(3) Bored or driven length (driving or boring cost); with a driven pile the length driven could be less than the actual length of pile.

Brickwork (Class U) — facework is not measured 'extra over' and this makes price assessment more difficult, since it involves the deduction of the common bricks to be replaced by facing bricks.

Measurement Processes

Dimensions Paper

All dimensions and mathematical calculations should be entered on separate sheets of dimensions paper or in dimensions books. These entries are to be carefully made so that they can be readily checked by another person without any possible chance of confusion arising.

The normal ruling of 'dimensions paper' on which the dimensions (scaled or taken direct from drawings) are entered, is indicated below.

1	2	3	4	1	2	3	4

Each dimension sheet is split into two identically ruled parts, each consisting of four columns. The purpose of each column will now be indicated for the benefit of those readers who are unfamiliar with the use of this type of paper.

Column 1 is termed the 'timesing column' in which multiplying figures are entered when there is more than one of the particular item being measured.

Column 2 is termed the 'dimension column' in which the actual dimensions, as scaled or taken direct from the drawings, are entered. There may be one, two or three lines of dimensions in an item depending on whether it is linear, square or cubic.

Column 3 is termed the 'squaring column' in which the length, area or volume, obtained by multiplying together the figures in columns 1 and 2, is recorded, ready for transfer to the abstract or bill.

Column 4 is known as the 'description column' in which the written description of each item is entered. The right-hand side of this wider column is frequently used to accommodate preliminary calculations and other basic information needed in building up the dimensions and references to the location of the work, and is referred to as 'waste'.

In the worked examples that follow in succeeding chapters the reader will notice that one set of columns only is used on each dimension sheet with the remainder used for explanatory notes, but in practice both sets of columns will be used for 'taking-off'.

Spacing of Items

Ample space should be left between all items on the dimension sheets so that it is possible to follow the dimensions with ease and to enable any items, which may have been omitted when the dimensions were first taken-off, to be subsequently inserted, without cramping up the dimensions unduly. The cramping of dimensions is a common failing among examination candidates and will cause loss of marks.

Waste

The use of the right-hand side of the description column for preliminary calculations, build-up of lengths, explanatory notes and related matters should not be overlooked. All steps that have been taken in arriving at dimensions, no matter how elementary or trivial they may appear, should be entered in the waste section of the description column. Following this procedure will do much to prevent doubts and misunderstandings concerning dimensions arising at some future date.

Order of Dimensions

A constant order of entering dimensions must be maintained through-out, that is (1) length, (2) breadth or width, and (3) depth or height. In this way there can be no doubt as to the shape of the item being measured. When measuring a cubic item of concrete 10 m long, 5 m wide and 0.50 m deep, the entry in the dimension column would be as follows.

10.00	Provision of conc. — designed
5.00	mix grade C10, ct. to BS 12, 20
0.50	mm agg. to BS 882; min. ct.
———	content 250 kg/m^3.
	&
	Placing mass conc. in base,
	thickness: 300–500 mm.

It will be noted that dimensions are usually recorded in metres to two places of decimals with a dot between the metres and fractions and a line drawn across the dimension column under each set of figures. Where the dimensions apply to more than one descriptive item, a bracket should be inserted as illustrated.

Timesing

If there were three such items, then this dimension would be multiplied by three in the timesing column as shown below.

3/	10.00		Provision of conc. — designed mix grade C10, ct. to BS 12, 20 mm agg. to BS 882; min. ct. content 250 kg/m³.
	5.00		
	0.50		
	————		&
			Placing mass conc. in base, thickness: 300–500 mm.

If it was subsequently found that a fourth bed was to be provided, then a further one can be added in the timesing column by the process known as 'dotting on', as indicated below.

3/	10.00		Descriptions as previous items
1·/	5.00		
	0.50		

Where there are a number of units of the same item, all multiplying factors should appear in the timesing column. Taking, for instance, 30 rows of piles with 4 piles in each row, entries on the dimension sheet would be:

| 30/4/ | 1 | | Number of preformed conc. piles, grade C25, 300 × 300 mm, len. 10.6 m; m.s. drivg. heads & shoes. |
| | ———— | | |

| 30/4/ | 10.00 | | Depth driven. |
| | ———— | | |

Abbreviations

Many of the words entered in the description column are abbreviated to save space and time spent in entering the items by highly skilled technical staff. Many abbreviations have become almost standard and are of general application; for this reason a list of the more common abbreviations is given in appendix I. A considerable number of abbreviations are obtained merely by shortening the particular words, such as the use of 'fwk.' in place of 'formwork', 'rad.' for 'radius' and 'conc.' for 'concrete'. The author also believes that it would be permissible to omit the metric symbols in billed item descriptions, but has included them in the examples, following the practice adopted in *The CESMM3 Handbook*.[5]

Grouping of Dimensions

Where more than one set of dimensions relates to the same description, the dimensions should be suitably bracketed so that this shall be made perfectly clear. The following example illustrates this point.

148.00	Clay pipes to BS 65 w.s. & s
———	flex. jts. nom. bore 225 mm in
246.00	trs, between mhs 8 & 12,
———	depth: 2–2.5 m.
132.00	
———	
56.00	
———	

Where the same dimensions apply to more than one item, the best procedure is to segregate each of the separate descriptions by an '&' sign as illustrated below, and to insert the customary bracket.

260.00	Excavn. for cuttgs. max. depth
16.00	2–5 m; Excvtd. Surf. 0.30 m
3.20	above Final Surf.
———	&
	Fillg. embankts. selected
	excvtd. mat. other than topsoil
	or rock.

Deductions

After measuring an item of construction it is sometimes necessary to deduct for voids or openings in the main area or volume. This is normally performed by following the main item by a deduction item as shown in the following example.

11.80	Placg. mass conc. a.b. in	
10.35	ground slab; thickness	
0.20	150–300 mm.	
1.60	Ddt. ditto.	
1.45		
0.20		(opgs.)

Figured Dimensions

When 'taking-off' it is most desirable to use figured dimensions on the drawings in preference to scaling, since the drawings are almost invariably in the form of prints, which are not always true-to-scale. It is sometimes necessary to build up overall dimensions from a series of figured dimensions and this work is best set down in waste, on the right-hand side of the description column.

Numbering and Titles of Dimension Sheets

Each dimension sheet should be suitably headed with the title and section of the project at the head of each sheet and with each sheet numbered consecutively at the bottom. Some prefer to number each set of columns on each dimension sheet separately. The entering of page numbers on each dimension sheet ensures the early discovery of a missing sheet and that the sheets are in the correct sequence.

At the top of the first dimension sheet for each main section of the work should be entered a list of the drawings from which the measurements have been taken, with the precise drawing number of each contract drawing carefully recorded. A typical example of such a list follows.

NORTH CREAKE OUTFALL SEWER   SHEET NR 1

Drawings
NC/SEW/1/0A   (Layout Plan)
NC/SEW/1/5A   (Sewer Sections)
NC/SEW/1/6B   (Sewer Sections)
NC/SEW/1/7B   (Sewer Sections)
NC/SEW/1/12A   (Manhole Details)

The importance of listing the contract drawings from which the dimensions have been obtained in this way, is that in the event of changes being made to the work as originally planned resulting in the issue of amended drawings, it will clearly be seen that these changes occurred after the Bill of Quantities was prepared and that variations to the quantities can be expected.

It is good practice to punch all dimension sheets at their top left-hand corner and fasten them together with Treasury tags.

Take Off Lists

Where the work contained in a project is complex or fragmented, it is good practice to prepare a take off list at the outset listing the main components in the order that they will be measured. This enables the person measuring the work to look at the project in its entirety, to reduce the risk of omission of items and to provide a checklist as the detailed measurement proceeds.

Query Sheets

It is good practice to enter any queries that may arise during the taking-off on technical matters on query sheets, which are normally subdivided vertically into two parts. The first column contains details of the matters on which clarification or amplification are required and the second column is used for the Engineer's reply. These sheets should always be prepared in the examination where appropriate.

Use of Schedules

When measuring a number of items with similar general characteristics but of varying components, it is often desirable to use schedules as a means of setting down all the relevant information in tabulated form. This assists with the taking off process and reducing the risk of error, and is particularly appropriate for the measurement of a considerable number of manholes as illustrated on pages 218 and 219.

Abstracting

When the items on the dimension sheets after squaring cannot conveniently be transferred direct to the appropriate section of the bill,
they may be grouped in an abstract, where they will be suitably classified and reduced to the recognised units of measurement preparatory
to transfer to the bill. The various phases of abstracting are described
and illustrated in chapter 16, where other bill preparation processes
are also described.

5 Measurement of Ground Investigation, Geotechnical Processes, Demolition and Site Clearance

Ground Investigation

Ground or site investigation may be part of a construction contract or be a separate contract wholly concerned with this activity.

The measurement of trial pits, trenches and boreholes is classified in CESMM3[1] B 1–3 **, amplified by some important additional description rules. Each group of pits, trenches and boreholes generates at least two bill items, namely the number of holes and the aggregate depth measured in metres, under various classifications. Separate items are needed where the work includes excavation in rock or other material, support to holes or backfilling, and removal of obstructions (measured by the hour). This approach enables costs that are proportional to the number of holes, such as moving boring rigs, to be kept separate from costs related to the depths of holes. Pumping of trial holes is given by the hour at a stated minimum extraction rate.

Rule A1 of Class B requires item descriptions for the number and depth of trial pits and trenches to state the minimum plan area at the bottom of the pit or trench or where the work is undertaken to locate services, the maximum length of the trench. While rule A2 requires item descriptions for the number and depth of trial pits and trenches (B 1 1–4 *) to identify separately those expressly required to be excavated by hand because of the higher costs entailed. The items for both light cable percussion and rotary drilled boreholes shall state the nominal diameter. The diameters of rotary drilled boreholes are related to core sizes. For instance a core size of 11 mm requires a hole size of 18 mm, and a core of 25 mm a hole of 32 mm; core sizes range from 11 to 150 mm.

Samples and tests are numbered with the descriptions, covering size, type and class in accordance with BS 5930 (*Code of practice for site investigations*). With samples a distinction is made between (1) 'undisturbed' (normally a solid core placed in an airtight tube) and 'disturbed' (loose excavated soil); (2) soft material and rock; (3) samples taken from 'trial pits or trenches or sources at surface' and those from 'boreholes'.

Rule C1 of Class B stipulates that items for ground investigation shall be deemed to include the preparation and submission of the records and results but not their analysis (rule M2).

Examples of typical ground investigation bill items follow.

Number	Item Description	Unit	Quantity	Rate	Amount £	p
	GROUND INVESTIGATION					
	Trial pits					
B114	Number in material other than rock, maximum depth 3–5 m; minimum plan area at bottom of pit: 2.25 m^2.	nr	34			
B130	Depth in material other than rock; minimum plan area at bottom of pit: 2.25 m^2.	m	136			
B160	Depth backfilled with excavated material.	m	136			
B170	Removal of obstructions.	h	15			
B180	Pumping at a minimum extraction rate of 8000 l/h.	h	30			
	Rotary drilled boreholes (nominal minimum core diameter: 100 mm)					
B310	Number.	nr	14			
B342	Depth with core recovery in holes of maximum depth 5–10 m.	m	112			
B360	Depth backfilled with cement grout.	m	112			
B370	Core boxes, 3 m long.	nr	28			
	Samples					
B412	Disturbed samples of soft material from the surface or from trial pits: minimum 5 kg; Class 3.	nr	102			
B421	Open tube samples from boreholes; 100 mm diameter × 600 mm long, undisturbed sample; Class 1.	nr	56			

Note: The code numbers in the CESMM3[1] have been used as reference numbers for the bill items, but this approach is optional.

Geotechnical and Other Specialist Processes

Class C covers specialist work in changing the properties of soils and rocks and is usually carried out by sub-contractors. These specialist operations are likely to generate method-related charges for bringing plant to the site and associated activities.

The measurement of grouting is subdivided into four principal items — length and inclination of drilling, number of holes to be drilled, number of injections of grout, and amount of grout materials to be injected (C 1–3, 4 and 5 **). Drilling for grouting is classified according to inclination under five classifications (C 1 and 2, 1–5 **), and the description must include the diameter of the hole to be drilled (rule A1 of Class C). Grout mixture components and proportions should be stated in item descriptions, or alternatively reference may be made to the appropriate Specification clause(s).

The measurement of diaphragm walls (C6) has been kept relatively simple compared with the normal approach to the measurement of excavation and concrete. It recognises the specialist nature of the design, which is usually the responsibility of the Contractor. It is, however, necessary to establish clearly whether the cost of activities, such as the preparation of projecting reinforcement to receive capping beams and the disposal of excavated material, is to be included in the sub-contract or the main contract.

The measurement of ground anchors (C7) and sand, band and wick drains (C8) comprise both enumerated items to cover such activities as the moving of plant from one location to another, and linear items to cover the cost of materials that are proportional to length, such as tendons and sand.

Readers requiring more detailed information on the measurement of these specialist techniques are referred to *The CESMM3 Handbook*.[5]

Demolition and Site Clearance

Class D embraces the demolition and removal of objects above the original surface of the ground and tree roots below it. Other items below ground, such as basements and base slabs, will normally be measured under Class E.

The number and content of the measured items in this class are restricted, since it is assumed that tenderers will have to inspect the site to assess the working conditions, items to be removed and their likely saleable value.

General clearance embraces shrubs and trees with a girth not exceeding 500 mm (trunk measured at 1 m above ground), tree stumps not exceeding 150 mm diameter, hedges and undergrowth, and pipelines above ground not exceeding 100 mm nominal bore. It is measured in hectares (D1), with clear identification of the particular area, preferably delineated on a tender drawing. Larger trees and stumps are enumerated in their respective girth and diameter ranges (D 2 1–5 0 and D 3 1–3 0).

Buildings and other structures also need to be clearly identified, possibly by reference numbers or letters or names on a tender drawing. The bill description must include the predominant materials and the total volume above ground within the external faces of enclosing walls and roof. Items for demolition and site clearance shall be deemed to include disposal of the materials arising from the works (rule C1 of Class D). Separate items are required for any materials or components which are to remain the property of the Employer.

Demolition of pipelines above ground exceeding 100 mm nominal bore and those within buildings, exceeding 300 mm nominal bore, are measured by length in metres. Prices include the demolition and removal of supports (rule C4). Rule 5.14 applies to the dimension of the pipeline given in the demolition description in Example 1, as it constitutes a single dimension (225 mm) in the CESMM3 range of dimensions (100 to 300 mm), and could equally well be applied to the volume of the building.

Worked Example

An example follows covering the measurement of site clearance work.

SITE CLEARANCE DRAWING NO. 1

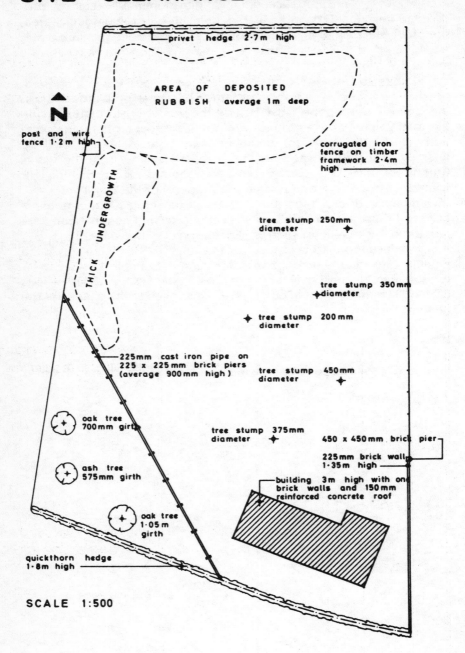

privet hedge 2·7m high

AREA OF DEPOSITED
RUBBISH average 1m deep

N

post and wire
fence 1·2 m high

corrugated iron
fence on timber
framework 2·4m
high

THICK UNDERGROWTH

tree stump 250mm
diameter

tree stump 350mm
diameter

tree stump 200mm
diameter

225mm cast iron pipe on
225 x 225mm brick piers
(average 900mm high)

tree stump 450mm
diameter

oak tree
700mm girth

tree stump 375mm
diameter

450 x 450mm brick pier

225mm brick wall
1·35m high

ash tree
575mm girth

building 3m high with one
brick walls and 150mm
reinforced concrete roof

oak tree
1·05m
girth

quickthorn hedge
1·8m high

SCALE 1:500

84

EXAMPLE I

Code numbers from CESMM3
inserted after item descriptions
for identification purposes.

78.00 45.00			Gen. clearance as Dwg. 1 inc. fences and hedges. D100	General clearance includes the demolition and clearance of all articles, objects and obstructions except buildings and other structures, and larger trees, stumps and pipelines. Area is reduced to hectares prior to billing.
2			Trees girth 500 mm – 1 m; holes back filld. w. excvtd. mat. D210	Felling of trees enumerated with girth classification as D2 1-5, and deemed to include removal of the stumps where they are also required to be removed (rule C3 of class D). The nature of any backfilling material is to be stated (rule A3 of class D).
1			Tree girth 1 – 2 m; ditto. D220	
5			Stumps diameter 150 – 500 mm. D310	All existing stumps of the same diameter classification as D310.

1.1

85

			bldg.- main area		
			18·500		
			7·000	129·500	
			projn.		
			6·000		
			2·000	12·000	Cubic contents calculated in 'waste' with each step suitably annotated.
			total area	141·500	
				3·000	
			total vol.	424·500 m³	
	Sum		Bldg. bwk. & r.c. roof vol. 250 - 500 m³ (425 m³).		Buildings to be demolished are entered as a sum with volume above ground level given in the description, in accordance with the ranges in D411-8. It would be more helpful to the contractor to be given the actual volume.
			D414		
			wall		
			23·75		
			0·23		
			1·35	7·36	
			pier		
			0·45		
			0·45		
			1·50	0·30	
				7·66	
	Sum		Bdy. wall, bwk.vol. 8 m³.		Dimensions are entered in the order of length, thickness and height. Distinction is made between a wall of which the volume can be given under D51, and linear fences included under general clearance. The actual volume is given as it is so far below 50 m³.
			D511		
	45·00		Pipeline nom. bore 225 mm.		Pipe supports are included in the price without the need for specific mention (rule C4 of class D).
			D610		

1.2

86

6   Measurement of Earthworks

Earthworks form a major part of most civil engineering contracts. The pricing of this work is made difficult by its relatively uncertain nature and extent and the effect of weather and water. The measurement rules attempt to recognise these factors and to permit the tenderer to make allowance for them.

The three measurement divisions in Class E are accompanied by extensive rules, which must be carefully considered and applied to every measured item. Excavation items cover excavation of one type (dredging, cuttings, foundations or general excavation) in one type of material (topsoil, rock, other natural material or artificial hard material); separate items are required for disposal of excavated material (disposal off site, disposal on site or re-use as fill). The items for foundations and general excavation are also classified according to the maximum depth range below the commencing surface. The depth stages are governed by various provisions — paragraphs 5.21 and 1.10–13 dealing with the definition and use of the various surface terms (original, final, commencing and excavated surfaces) and rule M5 of Class E with separation of items where separate stages of excavation are expressly prescribed.

For example, the depth of excavation for an item 'excavation for foundations' would normally be taken from the 'Original Surface' (surface of ground before any work has been carried out) to the 'Final Surface' (surface shown on the Drawings to which excavation is to be performed, to receive the Permanent Works). Where the depth of excavation in a measured item is to be restricted to a certain stage, this can be done by giving a 'Commencing Surface', where it differs from the 'Original Surface', at which it will begin or an 'Excavated Surface' at which it will end. Thus the stripping of topsoil or the excavation of the last 150–300 mm of a cutting, possibly left to protect the formation, may constitute separate and distinct stages of excavation.

The maximum depths listed in the third division of Class E for excavation of foundations and general excavation signify the total depth to be excavated, not the thickness of any layers of different material within it. In this way the bill rates can reflect the choice of plant and method of excavation.

However, excavation through a rock formation 3 m deep underlying 7 m of gravel would best be measured and described in two items: 'General excavation in material other than topsoil, rock or artificial hard material, maximum depth: 5–10 m; Excavated Surface top of the rock' and 'General excavation in rock, maximum depth: 2–5 m; Commencing Surface top of rock'.

Rule D1 of Class E prescribes that where material is not defined in the bill, it is deemed to be normally occurring soft natural material (other than topsoil, rock or artificial hard material). Some special categories of excavated material such as running sand are not listed, as this operation is akin to dealing with groundwater and constitutes a Contractor's contractual obligation. Rule M8 prescribes that an isolated volume of artificial hard material or rock occurring within other material to be excavated shall not be measured separately unless its volume exceeds 1 m^3, except that the minimum volume shall be 0.25 m^3 where the net width of excavation is less than 2 m.

Rule C1 provides that excavation items are deemed to include additional excavation to provide working space, upholding the sides of excavations and removal of dead services. Tenderers should allow for these items in excavation rates or in method-related charges. Excavation below a body of open water, such as a river, stream, canal, lake or body of tidal water, is measured separately in accordance with rule M7, at the highest applicable water level.

Rule M4 indicates that dredging is normally measured from soundings taken before and after the work is done. Where hopper or barge measurements are permissible this must be stated in a preamble to the Bill of Quantities, clearly identifying the circumstances in which the alternative method can be adopted. Dredging to remove silt is measured only where it is expressly required that silt which accumulates after the final surface has been reached shall be removed (rule M14).

Extra payment for double handling of excavated material is limited to that expressly required by rule M13. If the Contractor stockpiles without being instructed to do so he will not be entitled to additional payment, even although it might have been difficult to avoid it, as with excavated material to be subsequently used as fill. Excavation within borrow pits is classed as general excavation and shall be the net volume measured for filling, and is deemed to include the removal and replacement of overburden and unsuitable materials.

The quantities of material excavated or used as filling are measured net using dimensions from the drawings, with no allowance for bulking, shrinkage or waste (rule M1), with the exception of additional filling resulting from settlement or penetration into underlying material in excess of 75 mm in depth (rule M18) — a difficult provision to apply in practice. Filling items must distinguish between filling to structures, forming embankments, filling in layers to a stated depth or thickness (such as in drainage blankets, topsoiling, pitching and bleaching), and general filling. Rule A11 prescribes that differing compaction requirements relating to the same filling material are to be given in item descriptions. Filling items are deemed to include compaction (rule C3). However, filling and compaction cannot be included in excavation rates. Filling material shall be deemed to be non-selected excavated material other than topsoil or rock, unless otherwise stated in item descriptions (rule D6).

Filling to structures would include such work as filling around and over covered concrete storage tanks. Backfill to working space is not, however, measurable (rule M16). In items of filling to a stated depth or thickness, the materials shall be identified and work within three inclination ranges (10°–45°, 45°–90° and vertical) stated in the item descriptions (rules A13 and A14).

Under rule D7 the Contractor may use excavated rock as filling where the Specification permits, but he will only be paid at the rates for filling with excavated rock in locations where this is expressly required. Rules M20 and M21 requiring the measurement of the volume of rock fill in transport vehicles at the place of deposition in the case of soft areas and below water are often difficult to implement in practice.

The volume of disposal of excavated material measured shall be the difference between the total net volume of excavation and the net volume used for filling (rule M12). Disposal of excavated material shall be deemed to be disposal off the site unless otherwise stated in item descriptions (rule D4), and where disposal on site is required, the location shall be stated in the item description (rule A9).

The trimming of permanently exposed surfaces and the preparation of surfaces to receive permanent works, mainly concrete, are measured to both excavation and filling (rules M10, M11, M22 and M23). Work to sloping surfaces will be classified in the three categories listed in rules A8 and A15.

Turfing and seeding are measured in square metres separately identifying work to surfaces at an angle exceeding 10° to the horizontal (rule A18). Plants, shrubs and trees are enumerated stating the species and size, while hedges are measured in metres stating the species, size and spacing, and distinguishing between single and double rows.

Worked Examples

Worked examples follow, covering the measurement of excavation and fill.

Measurement of Excavation and Filling

Various methods can be used to calculate the volume of excavation and/or filling required as part of civil engineering works. The method used is often largely determined by the type of work involved. Accuracy and speed of operation are the main factors to consider when selecting the method of approach.

When calculating the volumes of excavation and filling for cuttings and embankments to accommodate roads and railway tracks, Simpson's rule can often be used to advantage and a simple example follows to illustrate this approach.

Using Simpson's rule the areas at intermediate even cross-sections (nrs. 2, 4, 6, etc.) are each multiplied by 4, the areas at intermediate uneven cross-sections (nrs. 3, 5, 7, etc.) are each multiplied by 2 and the end cross-sections taken once only. The sum of these areas is multiplied by one-third of the distance between the cross-sections to give the total volume. To use this formula it is essential that the cross-sections are taken at the same fixed distance apart and that there is an odd number of cross-sections (even number of spaces between cross-sections).

For instance, taking a cutting to be excavated for a road, 300 m in length and 40 m in width, to an even gradient, with mean depths calculated at 50 m intervals as indicated below and side slopes 2 to 1, and assuming that stripping of topsoil has already been taken.

Cross-section	1	2	3	4	5	6	7
Mean depth (m)	4	10	16	20	18	12	6

The width at the top of the cutting can be found by taking the width at the base, that is, 40 m and adding 2/2/the depth to give the horizontal spread of the banks (the width of each bank being twice the depth with a side slope of 2 to 1).

In simpler cases involving three cross-sections only, the prismoidal formula may be used, whereby

$$\text{volume} = \frac{1}{6} \times \frac{\text{total}}{\text{length}} \times \left\{ \begin{array}{c} \text{area of} \\ \text{first section} \end{array} + \begin{array}{c} \text{4 times area of} \\ \text{middle section} \end{array} + \begin{array}{c} \text{area of last} \\ \text{section} \end{array} \right\}$$

Cross-section	Depth (m)	Width at Top of Cutting (m)	Mean Width (m)	Weighting
1	4	$40 + 4 \times 4 = 56$	$\dfrac{56 + 40}{2} = 48$	1
2	10	$40 + 4 \times 10 = 80$	$\dfrac{80 + 40}{2} = 60$	4
3	16	$40 + 4 \times 16 = 104$	$\dfrac{104 + 40}{2} = 72$	2
4	20	$40 + 4 \times 20 = 120$	$\dfrac{120 + 40}{2} = 80$	4
5	18	$40 + 4 \times 18 = 112$	$\dfrac{112 + 40}{2} = 76$	2
6	12	$40 + 4 \times 12 = 88$	$\dfrac{88 + 40}{2} = 64$	4
7	6	$40 + 4 \times 6 = 64$	$\dfrac{64 + 40}{2} = 52$	1

The dimensions can now be entered on dimensions paper in the manner shown on the following sheet.

	48·00		Excavn. for cuttgs.;	To avoid a great deal of
	4·00		Commcg. Surf. 0·15m (c.s.1	laborious and unnecessary
			below Original Surf.	labour in squaring, all
4/	60·00		cube x 1/3 /50·00	dimensions have been entered
	10·00		(c.s.2	as superficial items, to be
			E220	subsequently cubed by
2/	72·00			multiplying the sum of the
	16·00		(c.s.3	areas by 1/3 of the length
				between the cross sections.
4/	80·00			
	20·00		(c.s.4	
2/	76·00			Total weighting is 18 and
	18·00		(c.s.5	the number of 50 m long
				sections of excavation is 6,
4/	64·00			so that 6/18 or 1/3 of the
	12·00		(c.s.6	distance of 50 m must be the
				timesing factor required.
	52·00			
	6·00		(c.s.7	
				Material to be excavated is
				deemed to be naturally
				occuring soft natural
				material other than topsoil,
				rock or artificial hard
				material, unless otherwise
				stated in item descriptions
				(rule D1 of class E).

Example II (Drawing No. 2)

This example covers the measurement of the excavation and filling required for an area 72 m × 36 m with surrounding banks with side slopes of 2½ to 1. The whole of the area, excluding banks, is to be stripped of topsoil, which will mainly be used for soiling the banks to a depth of 150 mm.

The 150.000 contour line is first plotted on the plan since this represents the demarcation line between the excavation and filling. Intermediate points on the contour line are found by interpolating between known spot or ground levels. For instance, taking the two levels in the bottom left-hand corner (S.W.), the difference between the two adjacent spot levels is 150.860 − 149.285 = 1.575 m, and the distance of the 150.000 level point from the edge of the area is

$$\frac{0.715}{1.575} \times 12.000 = 5.450 \text{ m}$$

The method of working adopted for this example is to calculate the volumes of excavation and fill in the main area (that is, 72 m × 36 m) from calculated average depths and to follow with the volumes of the banks. This is the simplest and quickest method although there are many alternative techniques. The average depths of excavation and fill are most conveniently found by suitably weighting the depth at each point on the grid of levels, according to the area that it affects. Generally this involves taking the depths at the extreme corners of the area once, intermediate points on the boundary twice and all other intermediate points four times. The sum of the weighted depths is divided by the total number of weightings (number of squares × 4) to give the average weighted depth for the whole area.

An alternative is to calculate the cross-sectional area on each grid line, including the section of adjacent bank, and to weight the areas in accordance with Simpson's rule. The banks at the end of each area would have to be added to the volumes of excavation and fill respectively.

Schedules of depths and the dimensions of excavation and fill now follow.

EXCAVATION AND FILLING
DRAWING NO. 2

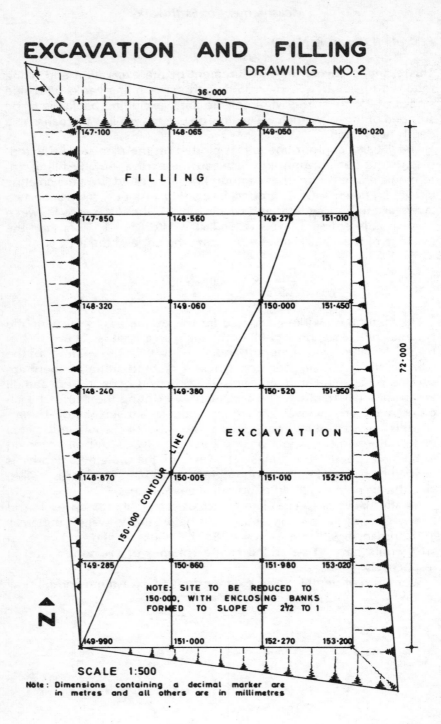

SCALE 1:500

Note : Dimensions containing a decimal marker are in metres and all others are in millimetres

94

AVERAGE DEPTH OF EXCAVATION TO MAIN AREA
(EXCLUDING BANKS)

GROUND LEVEL	DEPTH OF EXCAVATN.	WEIGHTING	WEIGHTED DEPTH OF EXCAVATN.	COMMENTS
150·020	0·150	1	0·150	150 mm topsoil
151·010	1·010	1	1·010	To weight this twice would give excessively high excavation quantities.
150·000	0·150	3	0·450	150 mm topsoil; affects 3 squares.
151·450	1·450	2	2·900	
150·520	0·520	3	1·560	
151·950	1·950	2	3·900	
150·005	0·150	3	0·450	150 mm topsoil
151·010	1·010	4	4·040	
152·210	2·210	2	4·420	
150·860	0·860	3	2·580	
151·980	1·980	4	7·920	
153·020	3·020	2	6·040	
149·990	0·150	1	0·150	150 mm topsoil
151·000	1·000	2	2·000	
152·270	2·270	2	4·540	
153·200	3·200	1	3·200	
		36	45·310	

Average depth of excavation
(including topsoil)   1·258

NOTE: The contour line is virtually coincident with the corners
of intermediate squares.
The total weighting of 36 is equivalent to 9 complete
squares with 4 effective levels to each.

2.1

EXCAVATION AND FILLING (Contd.)

AVERAGE DEPTH OF FILL TO MAIN AREA (EXCLUDING BANKS)

GROUND LEVEL	DEPTH OF FILL	WEIGHTING	WEIGHTED DEPTH OF FILL	COMMENTS
147·100	2·900	1	2·900	
148·065	1·935	2	3·870	
149·050	0·950	2	1·900	
150·020	—	1	—	negligible quantity
147·850	2·150	2	4·300	
148·560	1·440	4	5·760	
149·275	0·725	3	2·175	
148·320	1·680	2	3·360	
149·060	0·940	4	3·760	
150·000	—	3	—	
148·240	1·760	2	3·520	
149·380	0·620	3	1·860	
148·870	1·130	2	2·260	
150·005	—	3	—	negligible quantity
149·285	0·715	1	0·715	
149·990	0·010	1	0·010	
		36	36·390	

Average depth of fill	1·011	
Add replacement of topsoil	0·150	Much more convenient to add the additional 150 mm at the end rather than adding it to each individual depth.
Average total depth of fill	1·161	

2.2

			Excavation main area	Code numbers from CESMM3 inserted after item descriptions for identification purposes.

Excavation
main area

	72.00		Gen. excavn. topsoil, max.	Start with excavation of topsoil
	36.00		depth n.e. 0.25 m.	over main area.
	0.15		E411	Total volume of excavation to

Av. depth of excavn.   1.258
   less topsoil   150
                        1.108

Gen. excavn, max. depth
2-5 m; Comncg. Surf.
u/s of topsoil.

Code numbers from CESMM3 inserted after item descriptions for identification purposes.

Start with excavation of topsoil over main area.
Total volume of excavation to main area, using adjusted average depth previously calculated.
Dimensions are recorded to the nearest 10 millimetres.

½/ 72.00
    36.00
    1.11

E425

Fill
main area

½/ 72.00   Filling; gen., non-selected
    36.00   excvtd. mat., other than
    1.16   topsoil or rock.
                            E633

General excavated material to be used as filling, which is measured in m³ and is deemed to include compaction. The quantities of excavation and fill over the main area are reasonably balanced and so no item for disposal of excavated material has been taken.

½/ 72.00   Gen. excavn. topsoil, max.
    36.00   depth n.e. 0.25 m; Excvtd.
    0.15   Surf. u/s of topsoil.

This cubic item covers the stripping of topsoil over the area to be filled.

E411

2.3

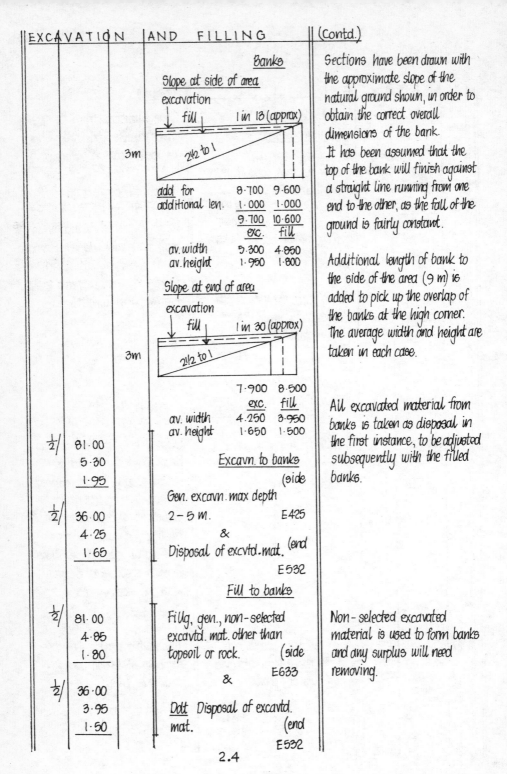

Banks

Slope at side of area

excavation

fill   1 in 13 (approx)

3m   2½ to 1

add for	8·700	9·600
additional len.	1·000	1·000
	9·700	10·600
	exc.	fill
av. width	5·300	4·850
av. height	1·950	1·800

Slope at end of area

excavation

fill   1 in 30 (approx)

3m   2½ to 1

	7·900	8·500
	exc.	fill
av. width	4·250	3·950
av. height	1·650	1·500

Excavn. to banks

(side

½/	81·00
	5·30
	1·95

Gen. excavn. max depth
2 - 5 m.   E425

½/	36·00
	4·25
	1·65

&

Disposal of excvtd. mat. (end

E532

Fill to banks

½/	81·00
	4·85
	1·80

Fillg, gen., non-selected
excavtd. mat. other than
topsoil or rock.   (side
                       E633

½/	36·00
	3·95
	1·50

&

Ddt Disposal of excavtd.
mat.   (end
                E532

Sections have been drawn with the approximate slope of the natural ground shown, in order to obtain the correct overall dimensions of the bank.
It has been assumed that the top of the bank will finish against a straight line running from one end to the other, as the fall of the ground is fairly constant.

Additional length of bank to the side of the area (9 m) is added to pick up the overlap of the banks at the high corner. The average width and height are taken in each case.

All excavated material from banks is taken as disposal in the first instance, to be adjusted subsequently with the filled banks.

Non-selected excavated material is used to form banks and any surplus will need removing.

2.4

EXCAVATION AND FILLING (Contd.)

<table>
<tr><td>½/</td><td>81·00
10·36</td><td rowspan="2">

<u>Bank slopes</u>

Trimmg. of excavtd. surfs., inclined at ∠ of 10° to 45° to hor.

            &   E512

Fillg. thickness 150 mm excvtd. topsoil; inclined at ∠ of 10° to 45° to hor.

            &   E641

Landscapg. grass seeding. to surfs. inclined at ∠ ex. 10° to hor.

                  E830

</td></tr>
<tr><td>½/</td><td>36·00
9·15</td></tr>
</table>

Average width of slope taken in each case for trimming, soiling and grass seeding. Trimming and filling to surfaces inclined at 10° or more to the horizontal are classified according to one of the three categories listed in rules A8 and A14, while seeding sloping surfaces are subject to the single classification of exceeding 10° (rule A18).

½/   81·00   Trimmg. of filled surfs.,
     9·45   inclined at ∠ of 10° to 45° to hor.

½/   36·00   &   E712
     8·50

Fillg. thickness 150 mm excvtd. topsoil; inclined at ∠ of 10° to 45° to hor.

            &   E641

Landscapg. grass seedg. to surfs. inclined at ∠ ex. 10° to hor.

                   E830

Trimming of slopes to excavation and filling are measured separately.

2.5

7 Measurement of Concrete

In Situ Concrete (Class F)

Class F in CESMM3 realistically prescribes separate bill items for the provision and placing of concrete. This separation is intended to assist the tenderer in relating prices more closely to costs and so simplifying estimating and assisting with the valuation of variations. A change of concrete mix or a variation in placing can be more readily accommodated by a change in the rate of one of the component items.

In measuring this class of work there will be a small number of items for the provision of concrete — one for each mix. These will be followed by many more items for placing concrete distinguishing between different types of concrete (mass, reinforced and prestressed), different locations and different structural members of different size ranges (second and third divisions). The prices for concrete can be inserted against three sets of items: (1) provision of concrete covering the materials and mixing costs; (2) placing of concrete embracing all labour and small tools in placing and curing concrete; and (3) method-related charges covering major plant and labour in batching and transporting concrete.

Provision of Concrete

The item descriptions for the provision of concrete use the terminology of BS 5328. *Methods for specifying concrete, including ready-mixed concrete.* Thus a 'standard mix' is a mix selected by the Contractor from a restricted list given in section 4 of BS 5328: Part 2. A 'designed' mix is one where the performance is stated in the contract and the mix proportions are selected by the Contractor to achieve the required performance. A 'prescribed' mix is one where the Engineer specifies the mix proportions.

Ordinary structural concrete (F 1–4) is normally made with Portland cement to BS 12, Portland blastfurnace cement to BS 146 or sulphate-resisting Portland cement to BS 4027, and must not contain any special additives. 'Standard mixes' vary between ST1 and ST5, while for 'designed mixes' commonly used grades are C7.5, C10, C15, C20, C25 and C30 or F3, F4 and F5. The maximum size of aggregate ranges between 10, 14, 20 and 40 mm.

A 'designed mix' (F2/F3) to comply with BS 5328 must encompass the particulars listed in the classification table, and also give the permitted size of aggregate, the minimum cement content in kg/m^3, the rate of sampling and possibly the water cement ratio and any quality assurance requirements. It is common practice for some of these requirements to be included in the specification rather than in item descriptions in the bill of quantities.

In the last type of mix (F4) — 'prescribed mix' — the item description should include the permitted type of aggregate, the mix proportions (in terms of weight of cement in kg), coarse aggregate, workability and any quality assurance requirements, and for fresh concrete the method of testing and rate of sampling. The rate of sampling is only required if it is higher than one sample for every 50 m^3 or 50 batches, whichever is the lesser volume. In both designated and prescribed mixes other optional statements can be made to elaborate the specification.[5]

Placing of Concrete

This activity is covered in CESMM3[1] by F 5–7 and the accompanying rules. The note at the foot of page 41 in the CESMM3[1] emphasises the effect of location on costs and the need to enlarge item descriptions in accordance with paragraph 5.10 where special characteristics affect the method and rate of placing concrete. Cost may be affected by height above or below ground, position and shape on plan, density of reinforcement, restrictions on access, unusual limitations on pouring, exceptional curing requirements and related aspects. The wording of 'may be stated' in the footnote is intentional — it should be implemented in the interests of tenderers, but it would not be satisfactory to make the provision compulsory, since an incorrect assessment of the likely differences in concrete placing costs could then be construed as an error in the bill.

The thicknesses of blinding bases, slabs and walls are expressed in ranges in the third division; thus all slabs exceeding 150 and not exceeding 300 mm in thickness can be grouped together in a single item. The quantity of concrete to be poured is measured in m^3, but small differences in thickness are not considered to be cost significant and hence the bill can be simplified by reducing the number of items in this way.

Tapering walls, columns and beams are best treated as non-standard components, stating the range of thicknesses or cross-sectional areas.

However, on occasions it may be sufficient to include the whole of the concrete placing in a single item where only a small proportion falls in a second and thicker category, provided that an additional description makes it clear what has been done and a preamble indicates a departure from CESMM3[1] in accordance with paragraph 5.4.

A nomogram in The CESMM3 Handbook[5] assists in determining the cross-sectional area of concrete components without calculation and so readily establishes the appropriate range in accordance with the third division particulars.

Columns and piers attached to a wall shall be measured as part of the wall and beams attached to a slab as part of the slab, except where required to be cast separately (rules M3 and M4), but the thickness of the wall or slab shall not be increased. On the other hand, concrete in suspended slabs and walls less than 1 m wide or long shall be measured as concrete in beams and columns respectively (rule D8). Some components may cause problems in classification, for example a column cap which might be classified as a thickening of the slab, thickening of column or 'other concrete form' (F 5–7 8). With these alternative approaches available an additional item description should be inserted to show what has been done in accordance with paragraph 5.13. The classification of 'other concrete forms' might also be used to cover composite members, giving the principal dimensions or an identifying reference (rule A4). Box culverts could be classified in this way where it would be more helpful to the tenderer than the separate measurement of walls and slabs.

Complex beam shapes make the pouring of concrete more difficult. Hence rules D9 and A3 require beams which are rectangular or approximately rectangular over less than four-fifths of their length, or where they are of box or other composite section, to be separately shown in the bill as 'special beam sections', with details of their cross-sectional dimensions or a drawing reference.

Rules M1 and M2 simplify the computation of concrete quantities by eliminating the need for deductions for reinforcement, prestressing components, and most cast in components, rebates, grooves, throats, chamfers, internal splays, pockets, holes, joints and the like, and the need for additions for small nibs or external splays. Internal splays arise from fillets placed inside formwork, thereby reducing the volume of concrete required, while external splays add to the volume of concrete. These rules result in a larger volume of concrete being measured than is actually required because significant voids are not adjusted while only small projections are ignored.

Cubic items for the placing of concrete are deemed to include the following additional items without the need for specific mention.

(1) Nibs or external splays less than 0.01 m² in cross-sectional area (rule M2 of Class F).
(2) Placing concrete against excavated surfaces (rule M2 (d) of Class G).
(3) Forming upper surfaces inclined at an angle not exceeding 15° to the horizontal (rule M3 of Class G).
(4) Formwork to temporary surfaces formed at the discretion of the Contractor (rule M2 (c) of Class G).
(5) Compacting concrete around reinforcement.

Concrete Ancillaries (Class G)

Formwork

The rules for the measurement of formwork are contained in G 1 ** to G 4 ** and the accompanying rules. Rules M1 and M2 provide guidelines to assist in deciding where formwork is measured. For instance formwork is to be measured to the surfaces of *in situ* concrete which require temporary support, except where otherwise stated in CESMM3.[1]
   The principal exceptions are:

(a) edges of blinding not exceeding 0.2 m wide;
(b) joints and associated rebates and grooves;
(c) temporary surfaces formed at the direction of the Contractor, such as surfaces of joints between pours;
(d) surfaces of concrete which are expressly required to be cast against excavated surfaces;
(e) surfaces of concrete which are cast against excavated surfaces inclined at an angle less than 45° to the horizontal.

   Formwork to upper surfaces shall be measured to surfaces inclined at an angle exceeding 15° to the horizontal and to other upper surfaces for which formwork is expressly required (rule M3).
   Rule D1 lists the angles of inclination to the vertical applicable to horizontal (85°–90°), sloping (10°–85°), battered (0°–10°) and vertical (0°) plane formwork.
   Formwork can be measured by length as a single item where concrete members or holes in them are of constant cross-section (note at bottom of page 43 of CESMM3[1]). Typical examples are walls, columns and beams, where the formwork can normally be re-used several times without major dismantling. An additional description will identify the members by their principal dimensions, mark number or other

reference. This is a good approach, since it highlights situations where many re-uses of formwork may be possible and enables the Contractor to price accordingly.

Where formwork is to be left in position for design purposes or through impossibility of removal, it becomes part of the Permanent Works and is to be measured in separately identifiable items (rule A1).

G1–4 classifies formwork according to the required finish, and it may be helpful to the tenderer to define rough and fair surfaces in a preamble. Surface features project beyond the surface. The classification by width in the third division permits a distinction to be made between areas of formwork generating differing ratios of labour cost to area. All narrow widths, fillets, splays and the like, not exceeding 200 mm in width, are measured as linear items. The widest category exceeds 1.22 m (full plywood panel width).

With curved formwork separate items are to be inserted for each different radius and each different shape of multi-radius formwork (spherical, conical, parabolic, ellipsoidal), desirably stating the location in each case (rule A4).

The rules for the measurement of formwork to small and large voids are each given in four stages of depth in the third division and the maximum diameters of circular voids and areas of other voids are given in rule D3.

The formwork to any void with a cross-sectional area exceeding 0.5 m^2 or diameter exceeding 0.7 m is treated as ordinary formwork (rule M4). No deduction is made for the volume of concrete displaced by a large or small void [rule M1(e) class F], even though the quantity can be considerable.

CESMM3[1] assumes that the tenderer will obtain much of the information needed to estimate the cost of formwork from the drawings, since, as described by Barnes[5] essential information such as support system, pour heights, striking procedure and number of re-uses cannot be described adequately in the bill. Full locational information should, however, be given in formwork item descriptions. The treatment of fixed and time-related formwork costs as method-related charges will assist in the valuation of variations.

The items of projections and intrusions in the third division are defined in rules D4 and D5.

Reinforcement

The measurement of reinforcement is detailed in classification G 5 **. Separate items are required for different reinforcement materials and sizes of the preferred dimensions listed in BS 4449, with bars of a diameter of 32 mm or more grouped together. Separate items are not required for tying and supporting reinforcement but the measured mass (weight) of steel reinforcement is to include supports to top reinforcement (rule C1).

Bars exceeding 12 m in length before bending are to be given separately in multiples of 3 m to give the tenderer the opportunity of allowing in his rates for the supplier's 'extra' for long bars and additional handling and fixing costs (rule A7). Where no length is stated in the bill description this signifies 12 m or less, while the inclusion of 15 m would indicate bars with lengths exceeding 12 m and not exceeding 15 m.

The descriptions of fabric reinforcement complying with BS 4483 need only give the fabric reference to BS 4483 or the wire and mesh arrangement in accordance with BS 4466, and the tenderer can obtain the relevant particulars from the British Standards. Additional fabric in laps is not measured (rule M9).

Joints

Movement joints are subdivided into two classes — those related to area (G 6 1–4 *) and those associated with length (G 6 5–7 *), while dowels are enumerated (G 6 8 *). The area items include formwork and the cost of supplying and fitting the filler materials. Linear items include waterstops and sealed rebates or grooves, and items for waterstops are deemed to include cutting and joining of waterstops and provision of special fittings at angles and junctions. Rule D8 distinguishes between 'open surface' joints, which are generally to horizontal surfaces requiring no stopend formwork, and 'formed surface' joints, which are normally of vertical surfaces with stopends. Joints are measured only at locations where joints are expressly required (rule M10), and they are all deemed to include formwork (rule C3).

Joint descriptions shall include full particulars of components such as thickness of filler materials and spacing and dimensions of dowel bars (rule A11). Rule M11 simplifies measurement by prescribing that all joint surface areas and width classifications shall be determined from the full width of the concrete member.

Post-tensioned Prestressing

Barnes[5] describes how the measurement of this work is kept simple on the basis that detailed specifications of stressing components and procedures are usually supplied and that the profiles and positions of stressing tendons can only be shown effectively on the Drawings. Rule A12 requires item descriptions to identify separately the different concrete members to be stressed and to include details of the components. Tenderers will have to build up their prestressing prices from an examination of the Drawings and Specification and their assessment of the quantities of ancillary work, such as ducts and vent pipes. The classification of tendons should be noted and that these are enumerated in prescribed stages of length (G 7 1–4 *).

Concrete Accessories

Finishes to *in situ* concrete, other than those resulting from casting against formwork, are measured in accordance with classification G8 1–2*. Items of top surface that are measurable include granolithic and similar finishes, stating the materials, thicknesses and surface finish of the applied layers (rule A13). The volume of these finishes shall not be included in the concrete measured in Class F.

Long inserts such as steel angles cast into concrete are measured in metres, while most other inserts, such as anchor bolts and pipe sleeves, are enumerated (G 8 3 1–2). Items for inserts are deemed to include their supply unless otherwise stated (rule C7). It is usually left to the Contractor to decide whether inserts should be cast or grouted in position. The item description should contain sufficient information to identify the work and there should be sufficient separate items to accommodate significant cost differences. For example, pipes passing through walls or slabs should be enumerated in ranges of nominal bore. Rule A15 distinguishes between inserts which project from one surface of the concrete, those which project from two surfaces of the concrete, and those which are totally within the concrete.

Precast Concrete (Class H)

The majority of precast concrete components are enumerated with the descriptions giving the position in the Works, specification of concrete and the mark or type number (rules A1 and A2). The tenderer will obtain the remainder of the information from the Drawings and Specification. The cost of the larger special components is influenced considerably by shape and size and the number required of each type; this information is given most effectively on drawings. Rule A2 prescribes that units (components) with different dimensions shall be given different mark or type numbers.

Hence rule A2 overrides the ranged classifications of length, area and mass listed in the second and third divisions of the classification table. Furthermore, paragraph 5.14 permits the use of specific dimensions in place of ranged dimensions in an item containing components with the same dimensions. The descriptions of the different precast concrete components (units) are built up in the following manner.

Precast Concrete Unit	Mark Number	Principal Dimensions of Cross-section	Cross-sectional Area	Average Thickness	Area	Length	Mass
Beams and columns	√	√				√	√
Slabs	√			√	√		√
Segmental units and units for subways, culverts and ducts	√	√					√
Copings, sills and weir blocks	√	√	√				√

Barnes[5] suggests that main mark numbers be used to distinguish different mould shapes, such as 51 and 52, with subdivisions of these to cover minor variations affecting length, positioning of holes or reinforcing details, such as 51/a and 51/b.

Rule D3 requires major concrete components cast adjacent to their final positions, such as railway bridge decks, to be measured as *in situ* concrete. However, large precast concrete bridge beams cast alongside the bridge and subsequently placed over each span are measured as precast units, since they involve the multiple use of formwork and the casting of the beams other than in their final position (rule D2).

Worked Examples

Worked examples follow covering the measurement of a mass concrete retaining wall, reinforced concrete pumping chamber and prestressed concrete beams.

MASS CONCRETE RETAINING WALL

DRAWING NO. 3

300 300

piers at 5m centres

mass concrete wall
(1 : 2 : 4)

4·000

continuous pocket of
ashes between piers

ground level

100 mm weep holes
at 1·8m centres

1·200

concrete foundations
(1 : 2½ : 5)

900

2·400

SECTION THROUGH WALL

1·200

2·400

300

300

750

P L A N

SCALE 1:50

108

For the purpose of this example a 30 m length of wall has been taken and the earthwork dimensions have been omitted.

Note: the principles adopted in this example would apply equally well to the measurement of reservoirs, settling tanks, bridge abutments, etc. built in concrete.

The code numbers in CESMM3 have been inserted after each item for identification purposes. They can also form the bill item references.

<u>In situ concrete</u>
<u>Provisn. of conc.</u>

30·00	Designed mix grade C7·5,
2·40	cement to BS 12, 20 mm
0·90	agg. to BS 882, min. ct.
	content 120 kg/m³
	F213

The measurement of concrete is subdivided into provision and placing.

The concrete mix may be 'standard','designed' or 'prescribed'. The use of the grades in section 2 of BS 5328 simplifies the approach.

```
                    av. thickness
         piers        of wall
      5) 30          1·200
      ──────          300
       6+1         2) 1·500
                      750
```

30·00	Designed mix grade C15,
0·75	ct. to BS. 12, 20mm agg.
4·00	to BS 882, min. ct. content
	180 kg/m³
7/ 0·75	F233
0·30	
4·00	(piers

Note the extensive use of abbreviations and the standard order of dimensions, i.e. length, breadth and height.
Reference can be made in the item description to sampling requirements as a specification clause.
Piers are taken at both ends of retaining wall.

<u>Placg. of conc.</u>

30·00	Mass bases, thickness:
2·40	ex. 500 mm.
0·90	F524

It is good practice to adopt the appropriate standard terminology (F52*).

3.1

ht. of 300 - 500 th. wall.

$$4.000 \times 2/9 = \begin{array}{r} 1.200 \\ 300 \\ 900 \\ \hline 0.89 \end{array}$$

av. thicknesses

500	300
1.200	500
2) 1.700	2) 800
850	400

ht. of wall ex. 500 th.

$$\begin{array}{r} 4.000 \\ 890 \\ \hline 3.110 \end{array}$$

30.00		Mass wall, thickness:
0.40		300 - 500 mm.
0.89		F543
7/ 0.75		
0.30		
0.89		(piers
30.00		Mass wall, thickness:
0.85		ex. 500 mm.
3.11		F544
7/ 0.75		
0.30		
3.11		(piers

Concrete Ancillaries

Fwk. fair finish

30.00	Slopg.	G225
4.10		

Note build up of dimensions in 'waste'.

The thickness of wall determines the amount of tamping or vibrating that has to be carried out for a given volume of concrete –this affects the price and the thickness is therefore classified in accordance with the ranges in the third division of Class F. The wall has to be subdivided into the part not exceeding 500 mm thick and that exceeding it.

Attached piers are included with the wall, in accordance with rule M3 of class F.

Assumed that concrete in wall foundation will be cast against excavated surfaces.

Wrot formwork has been taken for the full height of the wall as it would probably be difficult to use sawn formwork for the bottom section below ground only and it will avoid any snags arising from variations in the finished ground level. Formwork shall be deemed to be to plane areas and to exceed 1.22 m wide unless otherwise stated (rule D2 of class G).

3.2

<table>
<tr><td></td><td></td><td></td><td>less piers 30·000</td><td></td></tr>
<tr><td></td><td></td><td></td><td>7/750 5·250</td><td></td></tr>
<tr><td></td><td></td><td></td><td>24·750</td><td></td></tr>
</table>

		Fwk ro. finish.	
	24·75	Vert.	G145
	4·00		
7/	0·75	Vert. width 0·4 – 1·22 m.	
	4·00	(pier faces	
		G144	
7/2/	0·30	Vert. width 0·2 – 0·4 m.	
	4·00	(pier retns.	
		G143	

Note longer length of sloping face (scaled from drawing). Described as sloping and not battering as exceeds 10° from vertical.

The formwork to the faces and returns of the piers is kept separate from that to the wall face, as the narrow widths generate separate bill items. Both are superficial items as they exceed 200 mm wide.

1·800)30·000
17

17/	1	Inserts
		100 mm clayware land drain, 1 m lg., cast in on rake, totally within conc. vol. G832

Measured as inserts in accordance with G832. Separate items are not required for adapting formwork, as the inserts are not required to be grouted into preformed openings (rule M16 of class G).

Note: If expansion jointing was required between the various sections of wall, the non-extruding expansion jointing for the full cross-sectional area would be measured in square metres, with the strip of sealing compound on the outer face of the wall taken as a linear item.

	24·75	Ashes in continuous pocket behind wall.
	0·60	
	0·45	E618

3.3

PUMPING CHAMBER DRAWING NO. 4

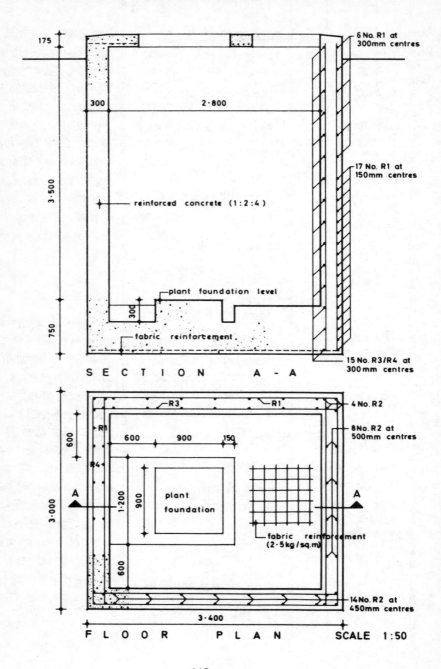

175

300   2·800

3·500

reinforced concrete (1:2:4)

6 No. R1 at
300mm centres

17 No. R1 at
150mm centres

plant foundation level

300

750

fabric reinforcement

15 No. R3/R4 at
300mm centres

S E C T I O N   A - A

R3   R1   4 No. R2

600   R1

600   900   150

8No. R2 at
500mm centres

A   R4

1·200   900

plant
foundation

fabric reinforcement
(2·5kg/sq.m)

A

3·000

600

14No. R2 at
450mm centres

3·400

F L O O R   P L A N   SCALE 1:50

112

PUMPING CHAMBER DRAWING NO.5

BAR SCHEDULE

Note: All bars are 12mm diameter

BAR REFERENCE	SHAPE OF BAR	LENGTH	TOTAL NUMBER
R 1	L SHAPED 1·650 / 1·800	3·450	92
R 2	STRAIGHT (in two lengths)	2·500	120
R 3	STRAIGHT	3·300	34
R 4	STRAIGHT	2·900	37
R 5	STRAIGHT	1·400	4
R 6	STRAIGHT	1·100	8
R 7	STRAIGHT	800	10
R 8	STRAIGHT	600	5

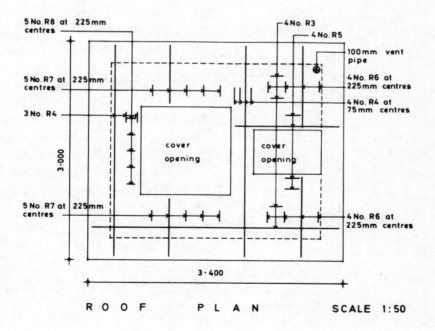

ROOF PLAN   SCALE 1:50

	total depth
	175
	3.500
	750
	4.425
less ht. above grd.	350
	4.075

Earthwks.

Gen. excavtn., max. depth:
2 - 5 m.

| 3.40 |
| 3.00 |
| 4.08 |

E425

Excavn. ancillaries

Prepn. of excvtd. surfs.

| 3.40 |
| 3.00 |

E522

Disposal of excvtd. mat.

| 3.40 |
| 3.00 |
| 4.08 |

E532

In situ concrete

Provsn. of conc. designed mix grade C15, ct. to BS12, 20 mm agg. to BS 882, min. ct. content 240 kg/m³

F223

| 2.80 |
| 2.40 |
| 0.68 |

	area		depth
less 2/300	3.400	3.000	less 750
	600	600	75
	2.800	2.400	675

base len.
600
900
150
1.650

Ddt. ditto.

| 1.65 |
| 1.20 |
| 0.23 |

(area ard.
plant. fdn.

4.1

Excavation for pits and similar structures is measured the total depth, but taken in the stages listed in the third division of Class E.
It is not considered necessary to separate the topsoil for subsequent re-use, due to the small quantity involved.

Separate items are not required for upholding sides of excavation or additional excavation to provide working space (rule C1 of class E), but disposal of excavated material requires measuring. Note: the principles of measurement adopted in this example would be equally applicable to reinforced concrete reservoirs, settling tanks, cooling towers, culverts, etc.

The provision and placing of concrete are measured separately as prescribed in Class F. It seems logical to take all cubic provision items first followed by the placing items in their various categories. Alternatively the activities could be taken simultaneously with each set of dimensions to avoid duplication.

		Provsn. of conc. a.b. F223	

0·90		Add ditto. (plant fdn.	Note method of building up the
0·90		ht.	girth of the chamber wall,
0·30		3·500	measured on its centre line, by
		750	taking the internal perimeter
		4·250	and adding the thickness of the
11·60		len.	wall for each corner.
0·30		2·800	
4·25		2·400	
	(walls 2/	5·200	The order of measurement
		10·400	follows a logical sequence –
3·40	add corners 4/300	1·200	the order of construction on the
3·00		11·600	site – base, walls and cover
0·18		(cover slab	slab.
1·20		Ddt. ditto.	Deduction of concrete is made
1·20			for the openings as they exceed
0·18			the area of 'large voids' as
			defined in rule D3 of class G.
0·90			
0·60			
0·18		(cover opgs.	

		Placg. of conc. Reinforced	
2·80		Bases and ground slabs,	Bases are measured in cubic
2·40		thickness : ex. 500 mm.	metres, distinguishing between
0·68		F624	different classes of concrete
			(mass, reinforced and prestressed)
			and separating into the thickness
1·65		Ddt. ditto. (area and	ranges listed in the third
1·20		(plant fdn.	division.
0·23			
			It is deemed desirable to keep
0·90		Small plant base,	the concrete in the plant base
0·90		thickness : 150 - 300 mm.	separate from the remainder
0·30		F622	because of its small volume.

4.2

		Placg. of conc. Reinfcd. a.b.	
11·60 0·30 4·25		Walls, thickness: 150 – 300 mm. F642	The walls are 300 mm thick and so fall within the thickness range of 150-300 mm.
3·40 3·00 0·18		Susp. slab thickness: 150 – 300 mm. F632	All suspended slabs are measured in cubic metres.
1·20 1·20 0·18		Ddt. ditto.	Same deductions for openings as before, as does not fall within the inclusion provisions in rule M1 of class F.
0·90 0·60 0·18		(cover (opgs.	

		Conc. ancillaries Fwk. fair fin: 2/1·650 3·300 1·200 4·500	Formwork providing rough and fair finishes must be distinguished and the plane classified in accordance with the second division of Class G (horizontal, sloping, battered, vertical and curved). Widths not exceeding 200 mm are measured as linear items and greater widths in square metres.
4·50 0·23		Vert. width: 0·2 – 0·4 m. (sump	
4/	0·90 0·30	(plant fdn. G243	

		Conc. Accessories	
2·80 2·40		Finishg. of top surfs., steel trowel. G812	To obtain smooth finish to concrete base.
4·50 0·23		Finishg. of formed surfs., steel trowel. G823	To vertical surfaces to sump and plant base.
3·60 0·30			

4.3

		Conc. ancillaries Fwk. ro. fin.	Note build up of external girth of pumping station.

		add 4/300 11·600
		         1·200
		        12·800
			Alternatively, the external dimensions of the chamber could be taken : 3·400

		less pt. 4·250
		above g.l. 225
		        4·025
			3·000
			      2/ 6·400
			        12·800

12·80
4·03
		Vertical (ext. face
		        (of walls
		
		        G145
			Unnecessary to state width as it exceeds 1·22 m (rule D2 of class G).

		Fwk. fair fin.	
12·80 0·23		Vertical width (ext. face 0·2 - 0·4 m. (of walls G243	Taking smooth face of concrete to 75 mm below ground level to allow for any irregularities in the finished ground surface. The formwork to the edges of the cover slab are taken later.

		3·500
		          75
		        3·575	

		11·600 less corners 1·200 10·400	

10·40
3·58
		Vertical (int. face
		        (of walls
		
		        G245
			Wrought formwork to internal faces of walls.

		Conc. accessories	
11·60 0·30		Finishg. of top slopg. surfs., steel trowel. G812	Sloping top surfaces to edges of cover slab. CESMM3 (class G) does not require the inclusion of the word 'sloping', but additional information may be given in accordance with 5.13 where advisable.
		Conc. Ancillaries Fwk. fair fin.	
2·80 2·40		Horizontal G215	Formwork to underside of cover slab.

Conc. ancillaries (contd.)

1·20	Ddt ditto.	
1·20		G215
0·90		(cover
0·60		opgs.

Formwork to underside of openings deducted as they exceed the large void areas prescribed in rule D3 of class G.

```
         1·200        900
         1·200        600
      2/ 2·400     2/ 1·500
         4·800        3·000
```

4·80	Vert. width :	(sides of
	0·1 – 0·2 m.	opgs.
3·00		G242

Linear items of formwork as not exceeding 200 mm wide.

12·80	Ditto.	(edges of
		cover slab
		G242

Conc. accessories
Inserts

1	100 mm dia. c.i. pipe proj. from one surf.	G832

The cover slab would be constructed later than the walls, after the plant has been installed – hence the need for a separate 150 mm strip of formwork to the edge of the cover slab.
Items for inserts shall be deemed to include their supply unless otherwise stated (rule C7 of class G).

Conc. ancillaries
Reinforcement

```
           3·400     3·000
   less 2/40  80        80
           3·320     2·920
```

3·32	High yield steel fabric to	
2·92	BS4483, nominal mass 2–3 kg/m², ref. A142.	
		G562

40 mm cover has been allowed to the fabric reinforcement to all edges in base slab.
Laps are not measured (rule M9 of class G).
Item descriptions for high yield steel fabric to BS.4483 shall state the appropriate reference (rule A9 of class G)

4.5

REINFORCED CONCRETE PUMPING CHAMBER (Contd.)

M.S. bars to BS 4449

	R3	R4
	34	37
less bars in cover slab	4	7
	30	30

92/	3·45	Diam. 12 mm (R1
120/	2·50	(R2
30/	3·30	(R3
30/	2·90	G514 (R4
		(walls
4/	3·30	Diam. 12 mm (R3
7/	2·90	(R4
4/	1·40	(R5
8/	1·10	(R6
10/	0·80	(R7
5/	0·60	G514 (R8
		(cover slab

Check the bar bending schedule against the Drawings before extracting the quantities from it. If no schedule is supplied, it will usually be necessary to prepare one.

40 mm cover is provided to the reinforcement unless otherwise specified and the normal allowance for hooked ends is an addition of 12 times the diameter of the bar for each hooked end. The total length of bar will be weighted up and billed in tonnes. Separate items are not required for supporting reinforcement (rule C1 of class G).

Bars exceeding 12 m in length are given separately in stages of 3 m (rule A7 of class G).

4.6

PRESTRESSED CONCRETE BEAMS
(SITE-MADE PRECAST POST-TENSIONED)

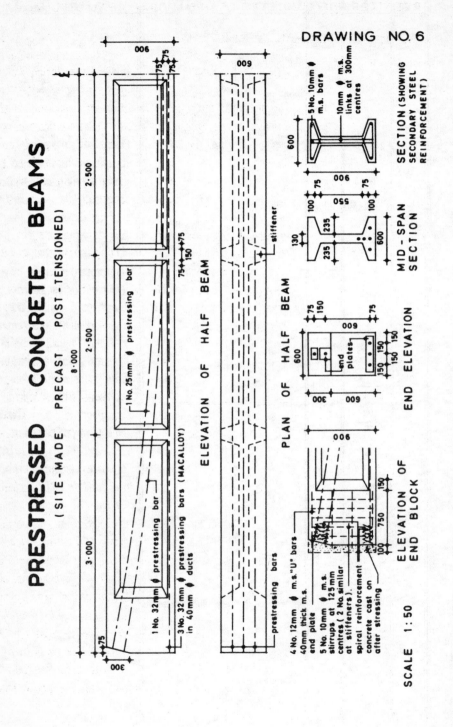

DRAWING NO. 6

ELEVATION OF HALF BEAM

1 No. 32mm ⌀ prestressing bar
3 No. 32 mm ⌀ prestressing bars (MACALLOY) in 40mm ⌀ ducts
1 No. 25mm ⌀ prestressing bar

SECTION (SHOWING SECONDARY STEEL REINFORCEMENT)

5 No. 10mm m.s. bars
10mm ⌀ m.s. links at 300mm centres

MID-SPAN SECTION

PLAN OF HALF BEAM

prestressing bars
stiffener

END ELEVATION

end plates

ELEVATION OF END BLOCK

prestressing bars
4 No. 12mm ⌀ m.s. "U" bars
40mm thick m.s. end plate
5 No. 10mm m.s. stirrups at 125mm centres (2 No. similar at stiffeners).
spiral reinforcement
concrete cast on after stressing

SCALE 1:50

120

(Site-made precast post-tensioned)			

PRECAST CONC.

<u>Designed mix for conc.</u>
<u>grade C 40 as Specfn.</u>
<u>clause — Bridge Deck</u>

40	Prestressed post-tensioned beams I sectn. 600 x 900 mm length: 16 m, mass : 12t., mark A1; prestressg. as Drawing Nr.6 and Specfn. clauses —

H357

Prestressed precast pre- and post-tensioned beams are enumerated, stating the length and mass ranges in accordance with the second and third division classifications in Class H. In this example the actual length and mass of each beam is stated as all 40 beams are identical.

Other particulars which need to be incorporated in the item description are:
(1) position in the works
(rule A1)
(2) specification of concrete
(rule A1)
(3) particulars of prestressing
(rule A3)
(4) cross section type
(rule A4)
(5) principal dimensions
(rule A4)
(6) mark or type number
(rule A2)

This approach necessitates adequately detailed drawings and comprehensive specification clauses. Concrete components which are cast other than in their final position shall generally be classed as precast concrete units (rule D2). The profiles and positions of stressing tendons can only be shown effectively on the Drawings.

5.1

8 Measurement of Brickwork, Blockwork, Masonry, Painting, Waterproofing and Simple Building Works incidental to Civil Engineering Works

This chapter covers several work sections from wall claddings to painting and waterproofing, with some common measurement principles running through them.

Brickwork, Blockwork and Masonry (Class U)

The CESMM3[1] rules for the measurement of brickwork, blockwork and masonry have to cover widely differing situations, ranging from wall cladding to small buildings, provided incidental to larger civil engineering works, to massive masonry bridge abutments, breakwaters and dock walls. Hence the second division classifications for brickwork range from half-brick walls to mass brickwork exceeding 1 m in thickness. Isolated walls having a length on plan not exceeding four times their thickness shall be classed as piers (rule D2 of Class U).

Similar approaches occur in the three categories of work with the measurement of walling, columns and piers, surface features and ancillaries. The materials have to be adequately described including the dimensions and types of the bricks, blocks or stones, and the type of bond, mortar, jointing and pointing (rules A1 and A3).

The actual thickness of brickwork or blockwork not exceeding 1 m thick shall be stated in the item description and the area is measured in m² (rule A5). Thicker walls are measured in m³ with the thickness stated. No deductions shall be made for holes or openings with a cross-sectional area not exceeding 0.25 m² (rule M2).

With walls of cavity or composite construction, each of the two skins shall be measured separately and suitably identified (rules M1 and A4). Wall ties across cavities and tying brickwork or blockwork to concrete, and concrete infills to cavities are each measured separately in m^2 (U18 5–6). The composite wall rules will also apply to brick walls built mainly of common bricks with facing bricks on the external face, and each will need to be measured separately with the average thickness stated. The third division classifies the work according to its general form such as vertical straight walls, vertical curved walls, battered straight walls, battered curved walls, vertical facing to concrete, battered facing to concrete and casing to metal sections.

Surface features such as rebates and band courses are measured as linear items with no additions to or deductions made from the main brickwork quantities (rules M2 and D3). The cross-sectional dimensions of surface features are included only where they exceed $0.05 \ m^2$ (rule A7). Item descriptions for surface features shall include sufficient particulars to identify special masonry and special or cut bricks and blocks, and to enable the estimator to calculate a realistic price for the work (rule A6). Sills and copings are measured as separate linear items and the areas and volumes of brickwork occupied by these features shall be excluded (rule M2). Fair facing is measured in m^2 regardless of the materials used to pick up the extra cost of working to a fair face, except for masonry where it is deemed to be included (rule C1 of Class U), although it would be much simpler to include the facework in the item description of the wall in all cases.

Item descriptions of damp-proof courses and joint reinforcement must include the materials and dimensions (rule A8). Building in of pipes and ducts are enumerated in two size ranges: the smallest covering pipes and ducts with a cross-sectional area not exceeding 0.05 m^2 and the largest exceeding $0.05 \ m^2$, stating the lengths in the descriptions where they exceed 1 m (rule A10).

Painting (Class V)

Painting carried out prior to delivery of components to the site is deemed to have been included in the items for the supply of the components, except off site treatment of structural metalwork in Class M. Item descriptions for painting shall include the materials and either the number of coats or the film thickness (rule A1 of Class V). Any work in preparing surfaces for painting is deemed to be included in the painting prices without need for special mention (rule C1).

The first division classifies the type of paint, the second division the type of surface and the third division different inclinations, restricted widths and isolated areas. Widths, not exceeding 1 m, are measured in two linear categories without distinguishing between different inclinations (rule M2). The enumerated isolated group provision can be applied only where the total surface area of each group does not exceed 6 m^2, and shall identify the work and state its location (rules D1 and A3). A typical example would be a coat of arms on wrought iron entrance gates.

No deductions are made for holes and openings in painted surfaces each not exceeding 0.5 m^2 in area (rule M1). In computing the painted area of metal sections, no allowance is made for connecting plates, brackets, rivets, bolts, nuts and the like (rule M4). The area of painted pipework surfaces is obtained by multiplying the pipe length by the barrel girth without deduction of flanges, valves and fittings; no additional allowances are made for the latter (rule C3) and the tenderer will need to cover these in his prices.

It will be noted that lower surfaces inclined at an angle not exceeding 60° to the horizontal are combined with soffit surfaces and that upper surfaces are subdivided into two categories. With types of surface, rough concrete is distinguished from smooth concrete because of the differences in labour and material requirements and resultant costs.

Waterproofing (Class W)

The rules for the measurement of waterproofing are similar to those for painting with comparable surface inclinations (excluding soffits but including curved and domed surfaces), restricted widths and isolated areas. Damp-proofing, tanking and roofing are each kept separate and item descriptions must include the material (classified in accordance with the second division) and the number and thickness of coatings or layers (rule A1 of Class W). Separate items are not required for preparation of surfaces, joints, overlaps, mitres, angles, fillets and built-up edges or for laying to falls and cambers (rule C1).

There is a similar deduction limit of 0.5 m^2 for holes or openings and of 6 m^2 for isolated groups of surfaces, as for painting. The classifications of curved or domed surfaces apply only where the radius of curvature is less than 10 m (rule D1).

Simple Building Works Incidental to Civil Engineering Works (Class Z)

Class Z of CESMM3 encompasses carpentry and joinery, insulation, windows, doors and glazing, surface finishes, linings and partitions, piped building services, ducted building services, and cabled building services. This is an entirely new section of CESMM3 and is intended to deal with simple building works which are incidental to civil engineering works, such as the pumphouse illustrated in example 1X. Work covered by classes contained elsewhere in CESMM3, such as drainage, metalwork, brickwork, blockwork and masonry, painting, asphalt work, and roof cladding and coverings, will be measured in accordance with the procedures prescribed in the other classes. More sophisticated building work is better measured in accordance with the Standard Method of Measurement of Building Works (SMM7).[8]

Carpentry and Joinery

Structural and carcassing timber is classified according to its location in the building, such as floors, walls and partitions, flat roofs, pitched roofs, plates and bearers and the like (Z 1 1 *), and measured in metres. Strip and sheet boarding are measured in m², stairs and walkways, and units and fittings enumerated and miscellaneous joinery, such as skirtings, architraves, trims and shelves separately measured as linear items.

When measuring lengths and areas for carpentry and joinery items, no allowance is made for joints or laps (rule M1 of Class Z), and no deduction for holes and openings each not exceeding 0.5 m² in area (rule M2). Sizes in item descriptions shall be nominal sizes unless otherwise stated (rule D1), while item descriptions shall state the materials used and identify whether sawn or wrought and any treatment, selection or protection for subsequent treatment (rule A1).

The item descriptions of most structural and carcassing timbers shall state the gross cross-sectional dimensions (rule A4), while those for stairs and walkways, and units and fittings, shall identify the shape, size and limits (rule A6).

Insulation

Insulation is classified according to type and location and measured in m² as Z 2 1–4 1–4, stating the materials and the overall nominal thickness (rule A7).

Windows, Doors and Glazing

Windows and doors are enumerated and classified according to the material and type of component as Z 3 1–3 1–6, which distinguishes between windows and window sub-frames and doors and frames or lining sets. Item descriptions for windows, doors and glazing shall indentify the shape, size and limits of the work (rule A8), and are deemed to include fixing, supply of fixing components and drilling or cutting of associated work (rule C2). Items of ironmongery are separately enumerated and described. Glazing is measured in m² with the item descriptions identifying the materials, nominal thicknesses, method of glazing and securing the glass (rule A10), while hermetically sealed units are separately itemised (rule A11). Patent glazing is measured in m² and classified as to roofs, opening lights or vertical surfaces.

Surface Finishes, Linings and Partitions

The principal finishes are classified as *in situ* finishes, beds and backing; tiles; flexible sheet coverings; and dry partitions and linings as Z 4 1–4 * with separate items for floors, sloping upper surfaces, walls and soffits, all measured in m², except surfaces with a width not exceeding 1 m, which are measured in metres in two separate width classifications (not exceeding 300 mm and 300 mm–1 m). Suspended ceilings are measured in m² in three depth of suspension ranges, as amplified by rule A21, while bulkheads are taken in metres and access panels and fittings enumerated. Proprietary system partitions are taken as linear items and classified as solid, fully glazed and partially glazed, with door units enumerated (Z 4 7 1–4).

Items for surface finishes, linings and partitions are deemed to include fixing, supply of fixing components and drilling or cutting of associated work (rule C3), and also preparing surfaces, forming joints, mitres, angles, fillets, built-up edges and laying to cambers or falls (rule C4), while suspended ceilings include associated primary support systems and edge trims (rule C6). The materials, surface finish and finished thickness shall be included in item descriptions (rule A17).

Piped Building Services

Pipework and insulation are each measured separately in metres, while fittings, equipment and sanitary appliances and fittings are each enumerated (Z 5 **). Lengths of pipes are measured along their centre lines and include the lengths occupied by fittings (rule M6). Items for piped services are deemed to include fixing, supply of fixing components and commissioning (rules C7 and C8). Item descriptions shall include details of location or type (rule A22), and pipework descriptions shall give materials, joint types and nominal bores (rule A23).

Ducted Building Services

The linear items of ductwork distinguish between circular and rectangular ductwork and between straight and curved ducts, with fittings and equipment enumerated and classified as Z 6 **. Item descriptions shall include locational and other details as rules A26, A27 and A28.

Cabled Building Services

Cables, conduits, trunking, trays, earthing and bonding are measured in metres, with their associated fittings enumerated, with the cables classified according to their positioning as Z 7 1 1–5. Guidance on measuring the lengths of the various components is provided in rules M8, M9 and M10. Final circuits and equipment and fittings (box fittings in the case of conduits) are enumerated and classified as Z 7 7–8 *.

Items for cabled building services are deemed to include determining circuits, terminations and connections, providing draw wires and draw cables, cleaning trunking, ducts and trays and threading cables through sleeves; fixing and supply of fixing components; and commissioning; while items for conduits are deemed to include fittings other than box fittings (rules C11, C12, C13 and C14). Item descriptions shall include locational and other details as rules A29, A30 and A31.

Worked Examples

Worked examples follow covering the measurement of a tall brick chimney shaft, a deep brick manhole, a stone-faced sea wall and a pumphouse, and embrace other work sections apart from brickwork, blockwork and masonry.

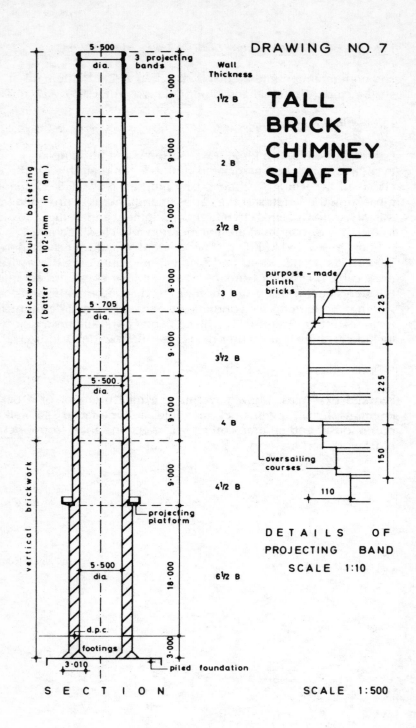

DRAWING NO. 7

TALL BRICK CHIMNEY SHAFT

5·500 dia.

3 projecting bands

Wall Thickness

1½ B

2 B

2½ B

3 B

5·705 dia.

3½ B

5·500 dia.

4 B

4½ B

projecting platform

5·500 dia.

6½ B

d.p.c.
footings

3·010

piled foundation

9·000
9·000
9·000
9·000
9·000
9·000
18·000
3·000

brickwork built battering
(batter of 102·5mm in 9m)

vertical brickwork

oversailing courses

purpose-made plinth bricks

225

225

150

110

DETAILS OF
PROJECTING BAND
SCALE 1:10

SECTION

SCALE 1:500

128

			TALL BRICK CHIMNEY SHAFT	EXAMPLE VI

Left page (dimension sheet):

(brickwork only measured)

BRICKWORK, BLOCKWORK & MASONRY

Eng. bwk. in class B
Southwater Reds to
BS 3921 in Eng. bond;
to chy. shaft.

Ftgs.

top cos. of ftgs.	1·505
bottom cos. of ftgs.	3·010
	2)4·515
av. thickness	2·258

int. diam.	5·500
2/½/wall thickness	1·453
2/½/ 1·453 (6½B)	
mean diam.	6·953

22/7/   6·95
   2·26
   0·98

Vert. curved wall, av.
thickness : 2·258 m, in c.m.
       U352.1

	3·000
less top of ftgs. to fdn. level	975
	2·025

22/7/   6·95
   1·45
   2·03

Vert. curved wall, nom.
thickness : 1·453 m, in c.m.
       U352.2
       (up to dpc.

Bwk. ancillaries

22/7/   6·95

Damp - proof course :
2 cos. of slates in c.m.
laid hor. to curve, width :
1·453 m.
       U382

Right page (notes):

Class U. heading.
Commence by measuring at the base and then work progressively up the chimney shaft.
The 13 courses of footings are averaged in width and measured in m³ as the thickness exceeds 1 m. All the preliminary calculations are entered in 'waste'. It is necessary to state the nominal thickness of the wall (rule A5). The wall thickness of 6½B comprises 6 x 225 (brick length + 10 mm joint) + 102·5 (½B) = 1452·5.
It is unnecessary to state that this brickwork is in footings.
The type of brick must be stated and also the form of construction. The diameter and height of the stack are not stated in the item description as CESMM3 assumes that tenderers will be supplied with fully dimensioned drawings.
The brickwork below dpc will be in cement mortar whilst that above may be in lime mortar to give flexibility and these must be kept separate. The change in wall thickness involves a change in suffix to the code reference.

Damp - proof courses are measured as linear items with a full description (rule A8 of class U).

6.1

			<u>above dpc</u>

$\frac{22}{7}$/ 6·95
1·45
<u>18·00</u>

Vert. curved wall, nom.
thickness : 1·453 m, in l.m.
U352.3

(0 – 18 m above g.l.

int. diam. 5·500
<u>add</u> 2/½/1·003 (4½B) <u>1·003</u>
mean diam. 6·503

The change in mortar involves a
further change in suffix in the
code reference.
The height stage is given in waste
for identification purposes.
Measurement of the engineering
brickwork continues up the chimney
shaft, leaving the fair face on the
exterior to be taken later.

$\frac{22}{7}$/ 6·50
9·00

Vert. curved wall, nom.
thickness : 1·003 m, in l.m.
U342.1

(18 – 27 m above g.l.

mean int. diam.
at 31·5 m   5·602
<u>add</u> 2/½/890 (4B)   <u>890</u>
mean diam. 6·492

This section of brickwork does
not exceed 1 m in thickness when
rounded off. The nominal
thickness has to be stated in
accordance with rule A5.
The thickness of 4½B consists of
4 x 225 + 102·5 = 1002·5 mm, and
the 4B wall is 3 x 225 + 215 =
890 mm.

$\frac{22}{7}$/ 6·49
9·00

Battered curved wall,
nom. thickness : 890 mm,
in l.m.   U344.1

(27 – 36 m above g.l.

mean int. diam. 5·602
<u>add</u> 2/½/778 (3½B) <u>778</u>
mean diam. 6·380

From this stage onwards the
brickwork is built battering and
this must be stated in the item
description as it involves further
additional expense, and as
prescribed in the third division.
The 3½B wall thickness is made
up of 3 x 225 + 102·5 = 777·5 mm.

The suffix to the code number
changes because of the different
thickness of wall.

$\frac{22}{7}$/ 6·38
9·00

Battered curved wall, nom.
thickness : 778 mm, in l.m.
U344.2

(36 – 45 m above g.l.

mean int. diam. 5·602
<u>add</u> 2/½/665 (3B) <u>665</u>
mean diam. 6·267

The 3B wall thickness consists
of 2 x 225 + 215, to incorporate
two internal 10 mm mortar joints,
and totals 665 mm.

6.2

$\frac{22}{7}$/	6.27		Battered curved wall, nom. thickness : 665 mm, in l.m.	All preliminary calculations should be recorded in 'waste' to
	9.00		U344.3	show the origins of dimensions and to permit checking.
			(45 – 54 m above g.l.	
			mean int. diam.　5.602	The thickness of the 2½B wall
			add 2/½/ 553 (2½B)　553	is made up of 2 × 225 + 102.5 =
			mean diam.　6.155	552.5 mm.
$\frac{22}{7}$/	6.16		Battered curved wall, nom. thickness : 553 mm, in l.m.	
	9.00		U344.4	
			(54 – 63 m above g.l.	
				Thickness of 2B wall
			mean int. diam.　5.602	comprises 225 + 215 = 440 mm.
			add 2/½/ 440 (2B)　440	The thickness of wall now
			mean diam.　6.042	drops below 500 mm and so there is a change to the code
$\frac{22}{7}$/	6.04		Battered curved wall, nom. thickness : 440 mm, in l.m.	reference.
	9.00		U334.1	
			(63 – 72 m above g.l.	Continue with mass engineering brickwork to top of shaft,
			mean int. diam.　5.602	ignoring for the moment the
			add 2/½/ 328 (1½B)　328	projecting bands.
			mean dia.　5.930	
$\frac{22}{7}$/	5.93		Battered curved wall, nom. thickness : 328 mm, in l.m.	The thickness of the 1½B wall consists of 225 + 102.5 = 327.5mm.
	9.00		U334.2	
			(72 – 81 m above g.l.	

6.3

131

		Surface features		
		Fair facing eng. bwk.		
		int. diam.	5.500	
		add 2 × wall-thickness		
		= 2/1.453	2.906	
		ext. diam.	8.406	

up to dpc

		below g.l.	75
		gl. to d.p.c.	150
			225

The detailed requirements of fair facing will be obtained from the Specification. The area is measured on the external surface of the shaft. If facing bricks were specified, it would be necessary to deduct the engineering brickwork for the volume occupied by the facing bricks (average 3/4 brick thick for English bond).

It is advisable to commence fair face 75 mm below ground level to allow for possible irregularities in the finished ground surface.

$\frac{22}{7}$/ 8.41 / 0.23 — Fair facg.; in c.m. U378·1

$\frac{22}{7}$/ 8.41 / 18.00 — Fair facg; in l.m. U378·2 (0 - 18 m above g.l.
5.500
add 2/1.003 2.006
7.506

$\frac{22}{7}$/ 7.51 / 9.00 — (18 - 27 m above g.l.
5.602
add 2/890 1.780
7.382

$\frac{22}{7}$/ 7.38 / 9.00 — (27 - 36 m above g.l.
5.602
add 2/778 1.556
7.158

$\frac{22}{7}$/ 7.16 / 9.00 — (36 - 45 m above g.l.
5.602
add 2/665 1.330
6.932

$\frac{22}{7}$/ 6.93 / 9.00 — (45 - 54 m above g.l.

6.4

The type of mortar will change at ground level, involving the use of suffixes in the code numbers. As these dimensions all relate to the same bill item, they can all be bracketted together and so eliminate the need to be continually repeating the description.

The waste dimensions are also abbreviated by omitting the repetitive descriptive notes.

The adjustments for the changing wall thickness can be taken from the waste dimensions for the main brickwork.


```
                        5·602
        add 2/553       1·106
                        6·708
```

22/7/	6·71 9·00	Fair facg. in l.m. U378·2 (54 - 63m above g.l.

Continue with fair facing dimensions to top of shaft. Next proceed to measure the three projecting bands at the top of the shaft. No adjustment is made to the main brickwork for these projecting features (rule M2). It is not necessary to include the cross-sectional area in the descriptions of these items as they do not exceed 0·05m² (rule A7).

```
                        5·602
        add 2/440        880
                        6·482
```

22/7/	6·48 9·00	(63 - 72m above g.l.

```
                        5·602
        add 2/328        656
                        6·258
```

22/7/	6·26 9·00	(72 - 81m above g.l.

```
                        6·258
        add 2/55         110
                        6·368
```

22/7/3/3/	6·37	Plinth bks; p.c.£52 per 100 in curved bands. U377 (78 - 81m above g.l.

Plinth brick items must include sufficient information for identification (rule A6).

```
                        6·258
        add 2/110        220
                        6·478
```

22/7/3/	6·48	Band courses; 2 cos. hi. on curve, total proj. of 110 mm. U374·1 78 - 81m above g.l.

Alternatively, these three projecting items might possibly be combined in a single item.

```
                        6·258
        add 2/55         110
                        6·368
```

In practice these measurements would be followed by the adjustment of brickwork for the inspection platform, building in of flue and test pipes, reinforcing steel bands and the like.

22/7/3/	6·37	Band courses; 3 oslg. cos. on curve, total proj. of 110 mm. U374·2 (78 - 81m above g.l.

6.5

DEEP BRICK MANHOLE
DRAWING NO. 8

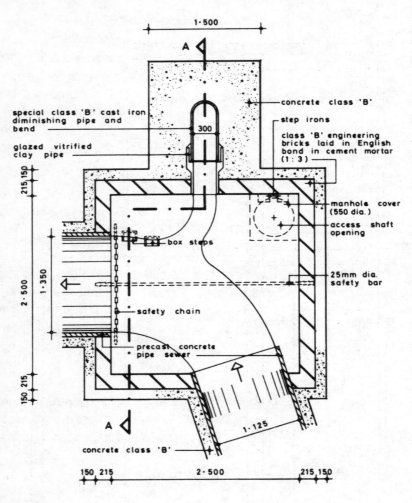

1·500

A

special class 'B' cast iron
diminishing pipe and
bend

300

glazed vitrified
clay pipe

concrete class 'B'

step irons

class 'B' engineering
bricks laid in English
bond in cement mortar
(1 : 3)

manhole cover
(550 dia.)

access shaft
opening

215 150

box steps

1·350

2·500

25mm dia.
safety bar

safety chain

precast concrete
pipe sewer

150 215

A

1·125

concrete class 'B'

150 215   2·500   215 150

NOTE
→ indicates direction of flow

P L A N

SCALE 1:50

134

DEEP BRICK MANHOLE
DRAWING NO. 9

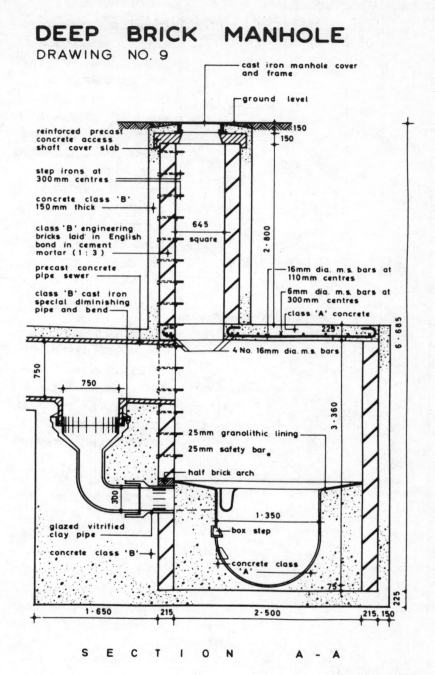

cast iron manhole cover and frame

ground level

reinforced precast concrete access shaft cover slab

step irons at 300mm centres

concrete class 'B' 150mm thick

class 'B' engineering bricks laid in English bond in cement mortar (1 : 3)

precast concrete pipe sewer

class 'B' cast iron special diminishing pipe and bend

150

150

645 square

2·800

16mm dia. m.s. bars at 110mm centres

6mm dia. m.s. bars at 300mm centres

class 'A' concrete

229

6·685

4 No. 16mm dia. m.s. bars

750

750

25mm granolithic lining

25mm safety bar

half brick arch

3·360

1·350

300

box step

glazed vitrified clay pipe

concrete class 'B'

concrete class 'A'

75

225

1·650   215   2·500   215 150

S E C T I O N   A - A

SCALE 1:50

135

Brickwork in manholes would normally be measured under Class K, in which case the whole of the work in the manhole would be a single enumerated item. This is a rather unusual and fairly complicated design of manhole and so is being measured in detail to illustrate the measurement of the component parts which straddle a number of work classes.

This approach is permitted by the footnote on page 53 of CESMM3.

Earthworks

```
                                2.500
        walls 2/215              430
        conc. 2/150             300
                              ──────
                               3.230
        total depth
  g.l. to top of base          6.685
                  base          225
                              ──────
                               6.910

   chamber depth      shaft depth
      6.910              2.800
      3.100               150
     ──────               150
      3.810             ──────
                        3.100
```

3.23	Gen. excavn;	(chbr.
3.23	max. depth: 5 – 10 m.	
6.91	&	E426
1.50	Fillg. to structures.	(backdrop
1.50		E613
6.91		

```
                              645
                2/215         430
                2/150         300
                            ──────
                             1.375
```

3.23	Add	
3.23	Disposal of	(chbr.
3.81	excvtd. mat.	
		E532
	&	
1.50		
1.50	Ddt	(backdrop
3.81	Fillg. to structures.	
		E613
1.38		(access
1.37		shaft
3.10		

Note method of building up dimensions in waste.
The whole of the excavation is taken as filling in the first instance as it will simplify the subsequent measurement of excavated material for disposal.

It is necessary to adjust soil disposal for the volume occupied by the chamber, backdrop and access shaft. Filling material shall be deemed to be non-selected excavated material other than topsoil or rock, unless otherwise stated in item descriptions (rule D6 of class E).

7.1

Excavn. ancillaries

3·23	Prepn. of excvtd. surfs.
3·23	E522
1·50	
1·50	

In situ conc.

3·23	Provsn. of conc.
3·23	designed mix grade C 10,
0·23	ct. to BS 12, 20 mm agg. to
	BS 882; min. ct. content
	210 kg/m³
	F 223

&

Placg. of conc. mass bases, thickness: 150 – 300 mm.
F522

1·50	Ddt. both last.
0·15	(surrd. to
0·23	(backdrop

1·65	Provsn. of conc. a.b.
1·50	F 223
2·60	&

Placg. of conc., mass surrd. to backdrop.
F580.1

Preparation of ground to receive base to manhole is measured in m². It shall be deemed to be carried out upon material other than topsoil, rock or artificial hard material, unless otherwise stated in item descriptions (rule D5 of class E).

As the dimensions for both the provision and the placing of concrete are identical, they are grouped together to avoid duplication of dimensions. As it is a designed mix, the minimum cement content has been included in the description.

Commence with the measurement of the concrete base and work up the manhole, following generally the sequence of construction.

This concrete does not slot in to any of the listed classifications in the second division and is therefore coded as 8 (other concrete forms). The depth of 2·60 is scaled off the drawing in the absence of relevant figured dimensions.

7.2

			len.
			2·500
			2·500
		2/	5·000
			10·000
	add walls 4/2/215		1·720
	conc. backg. 4/150		600
			12·320
			ht.
			225
			3·360
			75
			3·660

The concrete surround to the 750 mm diameter sewer has been left to be taken when measuring the sewer between manholes.

Note the method adopted for obtaining the perimeter length or girth of concrete backing to the chamber walls, measured on its centre line.

The dimensions follow the order of length, breadth and height.

12·32	Provsn. of conc. a.b. F223	
0·15	(chamber wall backg.	
3·66	&	

The placing has been classified as to walls, with the suffix 1 added at the end of the code to pick up 'backing'.

Placg. of conc., mass walls, thickness: n.e. 150 mm; backg.
F541.1

backdrop	pipe ext. diams.	
	int. bore	1·125
2·600	thickness 2/75	150
less base 225		1·275
2·375		
	int. bore	1·350
	thickness 2/85	170
		1·520

Adjustments need to be made for pipes exceeding 700 mm in external diameter (rule M1e of class F).

The thickness of pipe is added to the internal bore to give the external diameter.

1·50	Ddt. Provsn. of conc. a.b.	
0·15	F223	
2·38	& (backdrop	

The volume of backing concrete to be deducted for each pipe is πr²x thickness of backing.

22/7/	0·64	Ddt. Placg. of conc. F541.1
	0·64	mass walls, a.b.
	0·15	(1125 mm pipe

Adjustment has also to be made for the overlap of concrete backing to the chamber wall and the surround to the backdrop.

22/7/	0·76	
	0·76	
	0·15	(1350 mm pipe

The depth of 2·38 to the backdrop has been scaled in the absence of figured dimensions.

7.3

				len.	Build up of dimensions of
				645	concrete backing to access shaft
				645	walls.
			2/ 1·290	It is not considered necessary to
				2·580	make adjustment for the small
			add walls 4/2/215	1·720	thickness of concrete around the
			conc. backg. 4/150	600	access shaft cover slab.
				4·900	The order of length, breadth or
				ht.	thickness, and depth or height, in
				2·800	dimensions is maintained
				150	throughout for consistency.
				2·950	

4·90	Provsn. of conc. a.b. F223
0·15	& (access shaft
2·95	backg.

Placg. of conc. mass F541.1
walls a.b.

Conc. ancillaries.   Fwk.
len.
2·500
2·500
2/ 5·000
10·000
add walls 4/2/215   1·720
conc. backg. 4/2/150   1·200
12·920
ht.
225
3·360
300
3·885

12·92	Fwk. ro. fin. vert. G145
3·89	(chamber wall backg.
2/ 1·50	(sides to
2·60	backdrop surrd.

len.
645
645
2/ 1·290
2·580
add walls 4/2/215   1·720
conc. backg. 4/2/150   1·200
5·500

| 5·50 | Fwk. ro. fin. vert. G145 |
| 2·95 | (access shaft backg. |

The area of formwork represents the face of the concrete to be supported. It is measured to the side surfaces of in situ concrete cast within excavated volumes, except where the concrete is expressly required to be cast against excavated surfaces (rule M2d of class G).
The width of formwork is deemed to be exceeding 1·22 m wide unless otherwise stated (rule D2).

No deductions have been made to formwork for the large pipes nor has any cutting been measured.

Where the concrete will not be exposed, formwork producing a rough finish (sawn formwork) will be adequate.

7.4

	Cover slab
	2·500
2/215	430
	2·930

In situ conc.

2·93	<u>Provsn. of conc.</u> – designed mix grade C15, ct. to BS 12, 20 mm agg. to BS 882; min. ct. content 280 kg/m³.
2·93	
0·23	
	F233

The cover slab is reinforced and the concrete items will be followed by formwork and reinforcement.

No deduction of concrete is made for the access shaft opening as it does not exceed 0·5 m² (rule M1e of class F and rule D3 of class G).

&

<u>Placg. of conc.</u>, reinfd. susp. slab, thickness: 150–300 mm.

F632

Formwork to give a smooth finish to soffit of concrete cover slab to chamber.

<u>Conc. ancillaries</u>

2·50	Fwk. fair fin. hor.	F215
2·50		(cover slab soff.

The formwork to the edges of the slab exceed 200 mm wide and are therefore measured in m² under the appropriate third division classification.

4/	0·65	Fwk. fair fin. vert.
	0·23	width: 0·2–0·4 m. F243
		(edges to opg.

4/	2·93	Fwk. ro. fin. vert. width:
	0·23	0·2–0·4 m. (edges to slab
		F143

When calculating the length of bars, add 12 times the diameter of the bar for each hooked end. Divide the length of the slab by the spacing of the bars to give the number of spaces between them, and add one to total to convert the number of spaces into the number of bars.

	Reinft.
	len.
	2·930
less cover 2/40	80
	2·850
add hkd. ends 2/192	384
	3·234
	2·930
less cover	80
110)	2·850
	26+1

7.5

| | | DEEP BRICK MANHOLE (Contd.) | |

DEEP BRICK MANHOLE (Contd.)

		Reinforcement M.s. bars to BS 4449	Items for reinforcement shall be deemed to include supporting reinforcement other than steel supports to top reinforcement (rule C1 of class G).
27/	3·23	Diam. 16mm.　　G515	
4/	3·23	(transverse bars 110)645 　　　6	There are no hooked ends to distribution bars.
6/	0·64	Ddt. ditto.　　(opg. 　　　　　len. 　　　　　　2·500 less opg.　　645 　　　　　1·855 add len. over wall　215 　　　　　2·070 less cover 2/40　80 　　　　　1·990	Shorter bars finishing against access shaft opening.
6/	2·85	Diam. 6mm.　　G511	Number of bars obtained from Section A-A.
2/	1·99		
		In situ concrete	Benching is not included as a standard classification in the second division and it has therefore been included under 'other concrete forms', and comprises the second item in this category, and hence the inclusion of the second suffix.
	2·50	Provsn. of conc.- designed	
	2·50	mix grade C15, ct to BS12,	
	1·45	20 mm agg. to BS882; min. ct. content 280 kg/m³　F233	
		&	
		Placg. of conc. mass　F580.2 benchg. to m.h.　　channe. 　　　　　　1350 　　　　　1·125 　　　2)2·475 　　　　　1·238 add grano. ling. 2/25　50 mean diam.　2)1·288 　　　mean rad. 644	The average depth of the benching concrete has been scaled from the drawing.
½/22/7/	2·00	Ddt. both last	Length scaled off drawing measured on centre line of channel.
	0·64		
	0·64		It is not considered necessary to deduct for the 300 mm branch connection.
	2·00	(main	
	0·64	(channel	
	0·65		

7.6

141

Conc. Accessories

	2·50	
	2·50	Finishg. of formed surfs., grano. fin, steel trowel; as Spec. clause —. (sides of channel
2/	2·00	
	1·35	G823·1

Finishing benching with granolithic is measured in m², and is deemed to include materials, surface treatment, joints and formwork (rule C6 of class G).

BRICKWORK
Eng. bwk. class B to BS3921 in Eng. bond in c.m. (1:3).

```
                    gth. of chbr.
                        2·500
                        2·500
                    2/  5·000
                       10·000
      add corners 4/215  860
                       10·860
                gth. of access shaft
                         645
                         645
                    2/  1·290
                        2·580
      add corners 4/215  860
                        3·440
                    ht. of chbr.
                        3·360
                          75
                        3·435
```

Note build up of perimeter or girth of walls based on centre line measurements.
Item descriptions for brick walls shall either state the materials, nominal dimensions and types of brick or give equivalent references to applicable British Standard Specifications (rule A1 of class U).

	10·86	Vert. st. walls, nom. thickness:
	3·44	215 mm; to m.h. (chbr.
	3·44	(access shaft
	2·80	U321.1

One brick walls are measured in m². Additional descriptive items are separated from the standard descriptions by a semi-colon.

Bwk. ancillaries.

	1	Built-in pipe, csa 0·10 m²; w.h.b. ring over; supply of pipe m/s. (300 mm pipe
		U388·1

Built-in pipes are classified in two categories listed in the third division. In view of the large diameters of the pipes, the actual cross-sectional area is stated in each item description, measured to the outside of the pipe. The turning of a half brick ring over the pipe to prevent fracture is given as an additional non-standard description.

	1	Built-in pipe, csa 0·60 m²; w.h.b. ring over; supply of pipe m/s. (750 mm pipe
		U388·2

7.7

1		Built-in pipe, csa 1·29 m²; w.h.b. ring over; supply of pipe m/s. (1125mm pipe U388.3	Items for built-in pipes are deemed to include their supply unless otherwise stated (rule C2 of class U).
1		Built-in pipe, csa 1·83 m²; w.h.b. ring over; supply of pipe m/s. (1350 mm pipe U388.4	
22/7	0·44 0·44	Ddt. vert. st. wall, nom. thickness: 215 mm, to m.h. (750mm pipe	Brickwork is deducted for pipes whose external cross-sectional area exceeds 0·25 m² (external
22/7	0·64 0·64	(1125 mm pipe	diameter in excess of 564 mm) (rule M2 of class G).
22/7	0·76 0·76	(1350 mm pipe U321.1	Fair faced work is measured in m² on the actual face to be
	10·00 2·00	Fair facg. (chbr. U378	pointed. None is measured to brickwork covered by benching but the height above the benching
	2·58 2·80	(access shaft	is increased to 2 m to make allowance for the areas over pipes.
22/7	0·44 0·44	Ddt. fair facg. (750 mm pipe U378	
		Pipewk.	
	0·65	Clay pipe, BS class to BS 65 w. flex. jts., nom. bore 300 mm; in base of backdrop. I121.1	It is necessary to state the pipe material, nominal bore and jointing, and the appropriate British Standard reference (rule A2 of class I), and the location of
	1	Spun iron, class B to BS 4622 class 3, spec. comb. dimng. pipe & bend, nom. bore: 750-300 mm; in backdrop. J494.1	the pipe (rule A1 of class I). This is a special fitting but is measured in accordance with the rules in Class J.

7.8

DEEP BRICK MANHOLE (Contd.)

In situ concrete

22/7	0·60		
	0·45		
	0·45		

Ddt. provsn. of conc. (dimg. pipe
&   F223

22/7	1·50
	0·17
	0·17

Ddt. placg. of conc., mass surrd. to backdrop. (bend
F580.1

Deduction of concrete surround to backdrop for volume occupied by diminishing pipe and bend. The external radius is taken in each case.

Precast concrete
designed mix, grade C40

$$\begin{array}{rr} & 645 \\ \text{add walls } 2/215 & 430 \\ \text{o'hg. } 2/75 & 150 \\ \hline & 1·225 \end{array}$$

1

Precast reinfd. conc. slab, thickness : 200 mm ; area : 1·20 m²; mass : 500 kg - lt; recessed for opg. to m.h.   H523.1

Precast concrete slabs are enumerated giving the area and mass classification in accordance with class H. Items for precast concrete are deemed to include reinforcement, formwork, joints and finishes (rule C1 of class H).

Metalwork

1

C.i. m.h. cover & fr. to BS 497 ref. B4 - 2.   N999.1

2/ 1

C.i. boxsteps, mass 3·5 kg to m.h.   N999.2

15/ 1

W.i. step - iron, mass 1·75 kg to m.h.   N999.3

1

Galvd. w.i. safety chain, 1·50 m long, w. 2 nr. ragbolts b.i. conc. benchg. to m.h.   N999.4

1

Galvd. w.i. safety bar 25 mm diam., 2·50 m lg., w. 2 nr. ragbolts b.i. bwk. to m.h.   N999.5

These fittings are not specifically mentioned in CESMM3 and have each been enumerated and coded under class N (Miscellaneous Metalwork). Items for miscellaneous metalwork shall be deemed to include fixing to other work, supply of fixing components and drilling or cutting of other work (rule C1 of class N). Alternatively less information could be given in the bill description and the Contractor would then obtain the details from the Drawings and Specification in accordance with usual CESMM practice.

7.9

For layout reasons, and for ease of reading, this page has intentionally been left blank.

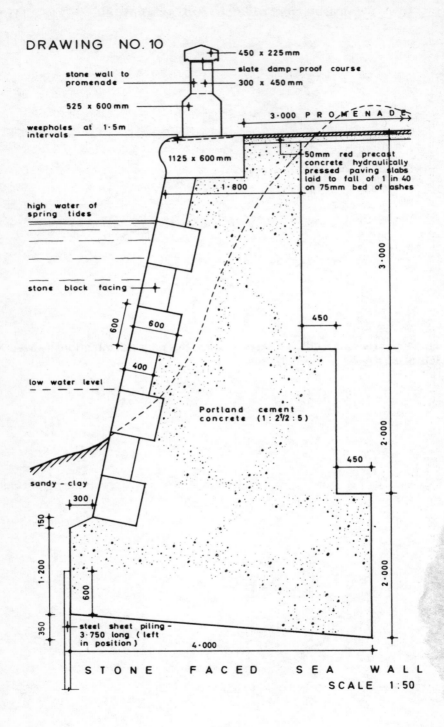

DRAWING NO. 10

stone wall to promenade

450 x 225mm
slate damp-proof course
300 x 450mm

525 x 600 mm

weepholes at 1·5m intervals

1125 x 600mm

3·000 P R O M E N A D E

50mm red precast concrete hydraulically pressed paving slabs laid to fall of 1 in 40 on 75mm bed of ashes

1·800

high water of spring tides

3·000

stone block facing

600

600

450

400

low water level

Portland cement concrete (1 : 2½ : 5)

450

2·000

sandy - clay

300

2·000

150

600

1·200

350

steel sheet piling - 3·750 long (left in position)

4·000

S T O N E   F A C E D   S E A   W A L L

SCALE 1:50

146

Note: The dimensions in this example have been squared ready for transfer to the abstract in Chapter 17.

Earthworks

	width
grd. surf.	300
hwl	1·650
2)	1·950
	975

60·00	Gen. excavn., max.	
0·98	depth: 1 – 2 m.	
1·55 91·1		E424

	depth
	1·800
	2·000
	2·000
	5·800

	width
hwl	2·150
excvtd. surf.	4·080
2)	6·230
	3·115

60·00	Gen. excavn., max.	
3·12	depth: 5–10 m;	
5·80 1085·8	Commg. hwl.	E426

Note method of obtaining mean widths by means of waste calculations or side casts. The first section is taken down to the high water level, as excavation below this level could be under water (the depth is scaled). The width of the second section includes the thickness of the steel sheet piling, and the front face of the excavation takes the form of a give and take line.

It is necessary to separate work to be carried out below high water level (rules M7 and A2 of class E). CESMM3 5.20 requires identification of the water levels in the preamble to the bill of quantities and on the drawings. Note the use of the terms 'Commencing Surface' and 'Excavated Surface' (CESMM3 5.21). The Commencing Surface shall be identified in the description of the work where it is other than the Original Surface, and the Excavated Surface requires identification if it is not the Final Surface (rule A4 of class E).

The Contractor will need to cover the cost of all temporary works in General Items or in his billed rates for earthworks, concrete and masonry.

Separate items are not required for upholding sides of excavation, additional excavation to provide working space and removal of dead services (rule C1 of class E).

8.1

Excavn. ancillaries

60.00		Disposal of excavtd.
0.98		mat.
1.55	91.1	E532
60.00		
3.12		
5.80	1085.8	
	1176.9	

Fillg. upper sectn.

	depth
	3.000
	250
	3.250
	width
	450
	450
	900

60.00		Ddt.
0.90		Disposal of excvtd.
3.25	175.5	mat. (rear upper sectn.
		& E532
60.00		
0.45		Add (rear lower sectn.
2.00	54.0	Fillg. to structures
		E613
60.00		
0.40		(frt. face
1.00	24.0	
	253.5	

8.2

It is necessary to measure the volume of excavated material which will be removed from the site and the quantity which will be required for filling to the void behind the sea wall.

The volume of disposal of excavated material is the difference between the total net volume of excavation and the net volume of excavated material used for filling (rule M12 of class E).

The disposal of excavated material is deemed to be disposal off site unless otherwise stated in item descriptions (rule D4).

The excavation quantities are taken as disposal and then adjustments made for the filling of the void.

Filling material is deemed to be non-selected excavated material other than topsoil or rock, unless otherwise stated (rule D6). Items for filling are deemed to include compaction (rule C3).

	STONE - FACED SEA WALL (Contd.)	

STONE - FACED SEA WALL (Contd.)

Excavn. ancillaries

60·00		
4·00	240·0	Prepn. of excvtd. surfs.
		E522

Piles

Interlocking steel piles, type 2N, section modulus 1150 cm³/m, grade 43A to BS 4360

60·00		Driven area.
3·75	225·0	P832

&

Area of piles of len: n.e. 14 m; treated w. 2cts. bit. paint. — P833

In situ concrete

```
         width
         2·750
         3·170
      2) 5·920
         2·960
```

60·00		Provsn. of conc. (base designed mix, grade
4·00		C10, ct. to BS 12, 40mm
2·00	480·0	agg. to BS 882. F224
60·00		& (middle sectn.
2·96		
2·00	355·2	

Placg. of conc. mass walls thickness: ex. 500 mm; backg. to masonry below h.wl. (upper sectn.

60·00		F544.1
2·10		
1·80	226·8	1062·0

Ddt. both last

½/	60·00		
	4·00		(tapg. base
	0·35	42·0	
	60·00		
	0·35		(top of toe
	0·35	7·4	49·4
		1012·6	8.3

The preparation of the excavated surface to receive the base of the wall is measured separately (rule M11).

Two items are required for the interlocking steel sheet piling – the driven area and the area classified according to the length categories in the third division of class P.

The concrete is subdivided into provision and placing and the placing is further subdivided into work above and below high water level in accordance with CESMM3 5·10 and the footnote on page 41 of CESMM3, whereby the location of concrete members in the works may be stated in item descriptions for placing of concrete where special characteristics may affect the method and rate of placing concrete. Some may feel that this subdivision of the work is unnecessary, provided the appropriate information with regard to water levels is shown on the drawing and referred to in the bill of quantities preambles.

The measurements include the volume occupied by the masonry which will need to be adjusted later. A suffix is added to the code reference on account of the additional description items.

It is often easier to measure the total enclosing rectangle and then deduct the voids from it.

		width	
		1·800	
		1·900	
		2)3·700	
		1·850	
60·00		Provsn. of conc. abd.	(upper sectn.
1·85			
0·60	66·6	&	F224
60·00		Placg. of conc. abd.	
0·75		above hwl.	
0·50	22·5		F544.2
	89·1		
		Conc. ancillaries	
		len.	
		2·000	
		2·000	
		1·800	
		5·800	
60·00		Fwk. ro. fin. vert;	
5·80	348·0	below hwl.	(rear face G145.1
60·00			
1·20	72·0		(frt. face
	420·0		
60·00		Fwk. ro. fin. slopg; width:	
0·35	21·0	0·2 - 0·4 m; below hwl.	
			G123.1

Finally, the concrete above high water level is measured.

It is unnecessary to repeat the previous descriptions when by using the expression abd (as before described), it is possible to refer back to the earlier items for the full descriptions.

Note the use of waste for preliminary calculations - recording in this way permits checking of the calculations as well as showing the build-up of dimensions.

A second suffix is added to the reference code for the placing item, because of the varied description.

The volume of the coping stone has been omitted from the dimensions. The concrete items are followed with the supporting formwork.

It is assumed that no formwork will be required to the stone face (the masonry serving as the formwork).

Formwork is deemed to be to plane areas exceeding 1·22 m unless otherwise stated (rule D2). Formwork to horizontal, sloping, battered and vertical surfaces are each kept separate and widths exceeding 200 mm are measured in m^2.

8.4

			600
			500
			1·100
60·00		Fwk. ro. fin. vert. width:	
1·10	66·0	0·4 – 1·22 m; above hwl.	
		G144.1	

It is considered desirable to state the width range in this case as the width does not exceed 1·22 m, even although it is part of a larger area in the same plane. However, some may feel that this separation is unnecessary.

Masonry

Ashlar masonry, P. st., flush ptd. w. mortar type M3

		av. width
4/600	2·400
3/400	1·200
7)3·600
514

The masonry is measured in m^2 as battered facing to concrete and stating the nominal thickness as rule A1 of class U.

The masonry has been subdivided between that above and below high water level, but the use of these classifications could be regarded as optional.

Items for masonry are deemed to include fair facing (rule C1).

60·00		Battered fcg. to conc;
4·20	252·0	nom. 600 & 400 in alt.
		cos., av. 514 mm, below hwl.
		U746.1

60·00		Ditto; thickness; nom.
0·70	42·0	400 mm above hwl.
		U736.1

The cutting to form the splayed top edge is not measured separately and the Contractor must allow for cutting in the masonry rates.

In situ concrete

60·00		Ddt. provsn. of conc. abd.
0·51		F224
4·20	128·5	&
		Ddt. placg. of conc. abd.
		below hwl. F544.1

Deduction of concrete (provision and placing) for the volume occupied by the masonry.

60·00		Ddt. provsn. of conc. abd.	
0·40			F224
0·70	16·8	&	
		Ddt. placg. of conc.	
		abd. above hwl. F544.2	

The use of code numbers will assist in identifying the items and quantities from which the deductions are to be made, and permit abbreviated descriptions.

		Ashlar Masonry
		Portland st. flush ptd.,
		mortar type M3.
60·00	60·0	Copg. 1125 x 600 mm
		rdd. w. sinkgs. as Dwg. 10.
		U771·1
60·00	60·0	Plinth 525 x 600 mm
		2ᶜᵉ splyd. as Dwg. 10.
		U777.1
60·00	27·0	Vert. st. wall; thickness:
0·45		300 mm, fair faced b.s.
		U731
60·00	60·0	Copg. 450 x 225 mm
		2ᶜᵉ wethd. & 2ᶜᵉ thro.
		as Dwg. 10.
		U771·2

The copings and plinths are measured as linear surface features, stating the cross - sectional dimensions, where the area exceeds 0·05 m², and labours: the latter cannot easily be related to cubic content (rule A7 of class U).
Reference to a drawing is usually necessary to provide adequate information for pricing.

The promenade wall is measured in m² stating the thickness and including fair face on both sides.

The coping stones may be connected by cramps and the plinth stones by dowels but these are not measured separately.

8.6

60.00	60.0	Dpc; width: 300 mm of 2 cos. of slates, ld. 6kg. jt. in mortar type M3. U782	Damp-proof courses are covered in ancillaries and are measured as linear items stating the material and the dimensions (rule A8 of class U). Additional material in laps is not measured (rule M5).

Promenade

60.00 1.30 0.30	23.4	Gen. excavn. max. depth: 0.25 – 0.5 m. E422 & Disposal of excvtd. mat. E532	Even shallow excavation is measured by volume, stating the depth classification in the third division. This is followed by the appropriate disposal item.

Light duty Pavement

60.00 3.00	180.0	Red precast conc. flags to BS 7263 type D; thickness 50 mm. R782 & Ash base, depth: 75 mm. R713	The promenade pavement is measured in accordance with the rules for class R (Roads and Pavings), which requires the separation of the concrete flags and the underlying ash base.

8.7

PUMPHOUSE

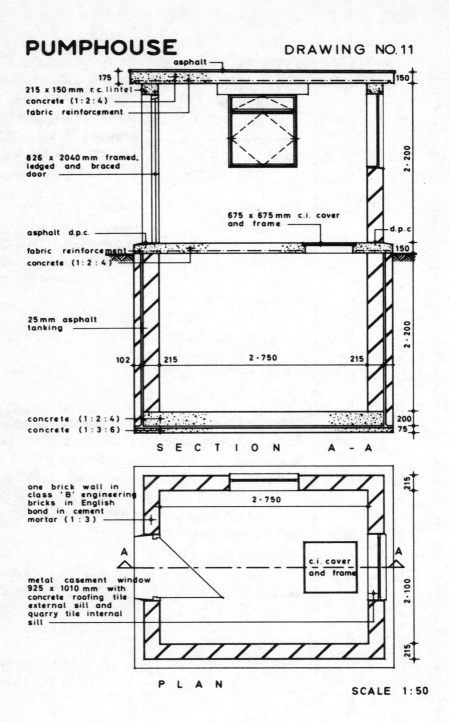

asphalt

175

215 x 150 mm r.c. lintel
concrete (1:2:4)
fabric reinforcement

150

826 x 2040 mm framed,
ledged and braced
door

2·200

675 x 675 mm c.i. cover
and frame

asphalt d.p.c.

d.p.c

fabric reinforcement
concrete (1:2:4)

150

25 mm asphalt
tanking

2·200

102   215   2·750   215

concrete (1:2:4)
concrete (1:3:6)

200
75

S E C T I O N   A - A

one brick wall in
class 'B' engineering
bricks in English
bond in cement
mortar (1:3)

215

2·750

A   A

c.i. cover
and frame

metal casement window
925 x 1010 mm with
concrete roofing tile
external sill and
quarry tile internal
sill

2·100

215

P L A N

SCALE 1:50

154

EXAMPLE IX

Buildings of this type are often encountered on civil engineering contracts, where the bulk of the work is civil engineering. Under CESMM3 the work is measured in accordance with class Z, which sets out the rules for the measurement of carpentry, joinery, finishes and services, while the other components are measured in accordance with the rules prescribed in the appropriate sections, such as earthworks, concrete, brickwork, pipework, painting and waterproofing work.

Preparation of excavated surface to receive concrete blinding.

The provision of insitu concrete is separated from the placing and measured in m³ with the particulars normally extracted from BS 5328.
Placing of blinding concrete is measured in m³; giving the appropriate thickness range. Formwork is not measured to the edges of the blinding concrete as they do not exceed 0·2m in width (rule M2a of class G).

Pumpwell

Earthworks

	len.	width
add	2·750	2·100
inner walls 2/215	430	430
asp. 2/25	50	50
outer walls 2/102	204	204
	3·434	2·784

	depth
	2·200
conc.slab	200
asp.	25
blindg. conc.	75
	2·500

3·43
2·78
2·50 — Gen. excavn., max. depth: 2–5m E425

&

Disposal of excavtd. mat. E532

3·43
2·78 — Prepn. of excvtd. surfs. E522

In situ concrete

3·43
2·78
0·08 — Provsn. of conc., designed mix grade C7·5, ct to BS12, 20mm agg. to BS882; min. ct. content 210kg/m³.
(blindg. F213

&

Placing of conc., mass blindg. thickness: n.e. 150mm. F511

9.1

<u>waterproofg.</u>

3.43		
2.78		

Tankg. of asp. to upper surfs. inclined at an ∠ n.e. 30° to hor; mastic asp. to BS 6577 in 2 coatgs: thickness: 25 mm. W211

The item description shall state the materials used and the number and thickness of coatings or layers (rule A1 of class W).
Separate items are not required for angle fillets and the like (rule C1).

<u>In situ concrete</u>

	len.	width
	2.750	2.100
add walls 2/215	430	430
	3.180	2.530

3.18
2.53
0.20

Provsn. of conc., designed mix grade C15, ct. to BS 12, 20 mm agg. to BS882; min. ct. content 280kg/m³ (base
F233

Follow with the in situ concrete base slab, comprising separate provision and placing items.

&

Placg. of conc. mass bases & grd. slabs; thickness: 150 – 300 mm.   F522

Select the appropriate work classification from the second division and the thickness range from the third division.

<u>Conc. accessories</u>

3.18
2.53

Finshg. of top surf., steel trowel.   G812

Finishing to top surface of the base slab to a smooth finish is measured in m² stating the method to be used.
It is considered unnecessary to deduct the area of finishing under the walls.
Note the method of obtaining the length of wall measured on its centre line.

<u>Brickwork</u>
<u>girth</u>

	2.750
	2.100
2/	4.850
	9.700
add corners 4/215	860
	10.560

9.2

PUMPHOUSE	(Contd.)	

Eng. bwk. Class B to BS 3921
in Eng. bond in c.m. (1:3).

| 10·56 2·20 | Vert. st. wall, nom. thickness: 215 mm, in composite wall to pumpg. chbr. U321.1 | Each skin of a composite wall shall be measured (rule M1 of class U. Some additional descriptive information has been given as provided for in CESMM3 5.13. |
| 9·70 2·20 | Fair facg. U378 | Internal fair face is measured to the surface in m². |

Waterproofg.
ht.
2·200
200
2·400
len.
10·560
add walls 4/215   860
asp.4/26   100
11·520

| 11·52 2·40 | Tankg; of asp. to surfs. inclined at an L ex. 60° to hor; mastic asp. to BS 6577 in 3 coatgs. thickness: 25 mm. W213 | Note build up of dimensions of asphalt in waste, starting with the girth previously calculated for the 215 mm brick wall on its centre line.

It is necessary to give both the thickness and the number of coatings of asphalt tanking, which can vary between horizontal and vertical coatings (see rule A1 of class W). The ranges of angles of inclination of the waterproofed surfaces are listed in the third division of class W. |

Brickwork
12·436
less 4/102   408
12·028

Eng. bwk. class B. to BS3921
in stretcher bond in c.m. (1:3)

| 12·03 2·40 | Vert. st. wall, nom. thickness: 102·5 mm, in composite wall to pumpg. chbr. U311.1 | Brickwork items shall include the types and nominal dimensions of bricks or refer to the appropriate British Standards (rule A1 of class U). |

9.3

			Superstructure	

Superstructure
Floor
In situ conc.

The leading dimensions are the same as for the blinding concrete at the base of the pumping chamber. The placing item description must include 'reinforced'. No deduction is made for the area (0·46 m²) occupied by the cast iron cover as it does not exceed the upper limit of a large void as prescribed in rule D3 of class G and provided for in rule M1e of class F.

3·43
2·78
0·15

Provsn. of conc. designed mix grade C15, ct. to BS 12, 20 mm agg. to BS 882; min. ct. content 240 kg/m³   F233

&

Placg. of conc. reinfd. susp. slab; thickness n.e. 150 mm.
F631

Conc. ancillaries
Fwk. fair fin.

2·75
2·10

Hor.   (u/s of flr. slab
G215

Formwork is deemed to be to plane areas and to exceed 1·22 m wide, unless otherwise stated (rule D2 of class G). The wall surfaces are excluded.

$$2/\frac{\begin{array}{r}3·434\\2·784\\\hline6·218\end{array}}{12·436}$$

12·44

Vert. width   (edges of slab
0·1 – 0·2 m.

Linear items as not exceeding 0·2 m wide.

(edges to cover opg.

4/   0·68

G242

Conc. accessories

It is not considered necessary to distinguish between the main horizontal surface and the sloping surfaces and the periphery. No deduction is made for the opening as the area does not exceed 0·5 m² (rule M15 of class G).

3·43
2·78

Finishg. of top surfs, steel trowel.   G812

9.4

		Conc. ancillaries		
		Reinft.		
		less	3·434	2·784
		cover 2/20	40	40
			3·394	2·744

3·39 2·74	Fabric high yield stl., to BS 4483, nom. mass 2–3 kg/m², ref. A142 G562

C.i. cover & fr.

| 1 | Cover & fr., c.i. to BS 497, ref. MC1R 60/60, set in. conc. flr. slab.   N999 |

Bwk.

Eng. bwk. class B to BS 3921 in Eng. bond in c.m. (1:3).

| 10·56 2·20 | Vert. st. wall, nom. thickness : 215 mm to superstructure of pumphse   U321.2 |

| 2/ | 10·56 2·20 | Fair fcg.   U378 |

---

20 mm cover has been allowed to the fabric reinforcement at the edges of the slab, but excluding the opening for which no deduction of reinforcement is made.

Item descriptions for high yield steel fabric to BS4483 shall state the fabric reference (rule A9 of class G).

No directions for measurement given in class N of CESMM3, so enumerated with a suitable description, including reference to the appropriate British Standard and coded as 4.5.

Girth of wall on its centre line calculated previously for 215 mm wall to pump well. The location of the work is included in the description to indicate that it is above ground in accordance with 5.13. Adjustment of brickwork for the door and windows will be made when measuring these components.

Fair facing is measured to both faces of the wall, with the centre line girth forming the mean of the inside and outside measurements. It appears to be the intention of CESMM3 to measure fair facing to brickwork, which consists of the pointing and careful handling of the bricks, as rule C1 of class U states that items for masonry shall be deemed to include fair facing, inferring that brickwork items do not. It would however seem more logical to describe the brick wall as built fair and pointed both sides.

10·56	Dpc, hor., width : 215 mm of asp. to BS 6577, 25 mm th. in 2 layers. U382	Damp-proof courses are measured as linear items irrespective of their width and the description is to include the material and the dimensions. The plane has been added under rule 5.13. The sub headings act as sign-posts and help others to find their way through the dimensions.

Roof

In situ conc.

	2·750	2·100
add		
walls 2/215	430	430
o'hg. 2/150	300	300
	3·480	2·830

3·48 2·83 0·16	Provsn. of conc. designed mix grade C15, ct. to BS 12, 20 mm agg. to BS 882; min. ct. content 240 kg/m³. F233 & Placg. of conc., reinfd. susp. slab; thickness : 150 – 300 mm. F632	The concrete is separated into the two items of provision and placing. In determining the thickness range for placing the concrete, the varying thickness of 150 to 175 mm has to be taken into account.

Conc. ancillaries
Fwk. fair fin.

2·75 2·10	Hor. (u/s of rf. slab G215	Unbroken area between walls of pumphouse finished to a smooth surface to give a satisfactory appearance.

	3·480
	2·830
2/	6·310
	12·620

12·62	Vert. width (edges of slab 0·1 – 0·2 m. G242 & Hor., width (eaves soff. 0·1 – 0·2 m. G212	Linear item as not exceeding 0·2 m wide, but distinguishing between those in horizontal and vertical planes. Forming the grooves on the soffit is not measured (rule M2b of class G).

9.6

<u>Conc. accessories</u>

3·48 2·83	Finishg. of top surf., stl. trowel. G812

Measured in m² stating the method to be used. It is not necessary to state that the surface is sloping.

<u>Conc. ancillaries</u>
<u>Reinft.</u>

less cover 2/20	3·480 40	2·830 40
	3·440	2·790

3·44 2·79	Fabric high yield stl. a.b.d. G562

Note insertion of basic calculations in waste as these form an important part of the taking off procedure. The letters 'abd' (as before described) eliminate the need for the repetition of full descriptions.

3·48 2·83	Roofg. asp. upper surfs. inclined at an ∠ n.e. 30° to the hor.; mastic asp. to BS 6577 in 2 coatgs., thickness: 25 mm. W311

The item description shall state the materials used and the number and thickness of coatings or layers (rule A1 of class W). This item is deemed to include preparing surfaces, forming joints, overlaps, mitres, angles, fillets, built-up edges and laying to falls or cambers (rule C1 of class W). Where protective base layers are required, these are measured separately as W4✳✳.

<u>Door</u>

1	Dr., wrot. swd., frd., ledged & braced, 826 x 2040 x 50 mm th., as specfn. clause ——. Z313

The description shall include the size of the door (rule A8 of class Z).

9.7

161

PUMPHOUSE					(Contd.)

Doors and ironmongery items are deemed to include fixing (rule C2 of class Z), and materials shall be stated in item descriptions for iron-mongery (rule A9 of class Z). Alternatively, doors and their associated frames and iron-mongery can be combined in single comprehensive items (footnote on p. 103 of CESMM3).

	2		Hinges, pressed stl. tee, 300 mm lg.
			Z341

	1		Latch, cylinder rim night, Yale 77.
			Z343

		add	826
		fr. less reb. 2/62	124
			950
		add	2·040
		fr. less reb.	62
			2·102

Build up of overall dimensions of door frame in waste.

	1		Dr. fr., wrot. swd., 112 × 75 mm, reb. × 2102 × 950 mm ov'll.
			Z314

The door frame is enumerated giving the sizes (rule A8 of class Z). Alternatively, the item description could refer to a drawing.

Painting is measured in accordance with class V. Painting to door surfaces with three coats of oil paint and including preparation work of knotting, priming and stopping (rules A1 and A2 of class V). No distinction is made in CESMM3 between painting internal and external surfaces. The timesing factor of 1½ makes allowance for the painting of edges to frames, ledges and braces of the door.

Paintg.

2/1½	0·83		K ps & ③ oil paint on tbr. surfs. inclined at an L ex. 60° to the hor.
	2.04		(dr. surfs.
			V323

9.8

162

					826	Build up of length on centre

PUMPHOU$E (Contd.)

```
                                          826
                      2/2·040            4·080
              add angles 2/75             150
                                        ─────
                                        5·056

5/      5·06      Kps & ③ oil paint on tbr.
                  surfs., width n.e. 300 mm.
                                  (dr. fr.
                                  V326
```
Build up of length on centre line of door frame to be painted in waste. The internal and external work is executed and measured separately, but is not distinguished in the item descriptions.

```
                      Wdws.

2/       1        Wdw., stl., hot dip galvsd.,
                  size 925 x 1010 mm to
                  BS 6510, as specfn. clause —.
                                  Z321
```
Metal windows are enumerated giving the size. The items are deemed to include fixing, supply of fixing components and drilling and cutting of associated work (rule C2 of class Z).

```
2/      0·90     Glazg., stand. plain glass,
        0·20     clear float, nom. thickness:
                 3 mm, to BS 952, to stl.
2/      0·90     wdws. fxd. w. met. putty
        0·75     & stl. clips.
                                  Z351
```
Glazing item descriptions shall state the materials, their nominal thickness and method of glazing and securing the glass (rule A10 of class Z).

```
                      Paintg.

2/2/    0·93     Prime & ③ oil paint to
        1·01     met. sectns. of wdws. (mesd.
                 pane area on b.s.).

                                  V370
```
The painting to the metal windows is taken as metal sections in m², although it is not clear from CESMM3 as to whether this is the intended approach. Additional information is given in accordance with 5.13. No distinction is made between internal and external painting.

```
                      9.9
```

| | | | Adjustment of dr. & wdw. opgs. | The brickwork and fair facing is deducted for the areas occupied by the door, windows and lintels, as they each exceed 0·25m² in area (rule M2 of class U). |

				dr.	wdws.
			add lintel	950	925
			beargs. ²/100	200	200
				1·150	1·125

The abbreviated 'ddt.' is used frequently for deduct, along with many other standard abbreviations listed in Appendix 1.
The location of each item is shown clearly in waste.

	0·95		Ddt. vert. st. wall, nom.
	2·10		thickness: 215 mm in (dr. & fr.
			eng. bwk. a.b. U321.2
2/	0·93		
	1·01		(wdws.
	1·15		
	0·15		(lintel to dr.
2/	1·13		
	0·15		(lintels to wdws.

2/	0·95		Ddt. fair facg.	The same areas, twice timesed, are deducted for fair facing, to cover both internal and external faces.
	2·10		(dr. & fr.	
			U378	
2/2/	0·93			
	1·01		(wdws.	
2/	1·15			
	0·15		(lintel to dr.	
2/2/	1·13			
	0·15		(lintels to wdws.	

(Contd.)

Precast r.c. lintels to mix of 1:2:4/20 mm agg. reinfd. w. 2nr. 12mm m.s. bars.

1   Lintel, 150 x 215 x 1150 mm lg. mass : n.e. 250 kg.

HIII.1

1   Lintel, 150 x 215 x 1125mm lg.; mass ; n.e. 250 kg.

HIII.2

The lintels have been measured in accordance with the rules in class H for precast concrete beams, as being the most appropriate item, and giving the specification of the concrete as rule A1 and the principal dimensions as rule A4 of class H. Precast items are deemed to include reinforcement, form-work, joints and finishes (rule C1 of class H).

2/ 2·10
0·10

2/2/2/ 1·01
0·10

Fair facg.   (dr. reveals

(wdw. reveals

U378

Fair face is measured to door and window reveals in m². The windows have reveals finished fair face both internally and externally, and hence the additional timesing factor of two.

Surf. finishes

2/ 0·93

Wdw. sill, ext.; 2 cos. of conc. rfg. tiles, 25mm total thickness ; width : 110 mm.

Z425.1

&

Ditto., int., clay quarry tiles, 18 mm th.; width: 150 mm.

Z425.2

The tile sills are measured by length as they do not exceed 300 mm wide, and the item descriptions shall include the material, surface finish where appropriate, and finished thickness (rule A17 of class Z).

9 Measurement of Piling

Piles (Class P)

Some piling activities are very appropriately covered by method-related charges. For example, the transport of piling plant and equipment to the site and the erection of stagings can be covered by fixed charges and the operation of piling plant by time-related charges.

The rules in Class P aim to obtain a set of prices for each piling operation, which will lead to equitable payment in the event of variations. The bill items relate to groups of piles which are piles of the same type, material and cross-section in a single location. On a small contract all the piles might be regarded as in a single group.

Two or three separate billed items are generated for each group of piles by the classifications in the third division, depending on the type of pile. They embrace the following elements:

(1) number of piles in a group to cover the plant and labour costs involved in moving the rig from one pile position to the next, setting up at each position and preparing to drive or bore;
(2) length of piles in the group, covering the material cost of the piles; the concreted lengths of cast in place piles shall be measured from the cut-off levels to the toe levels expressly required (rule M3);
(3) total depth bored or driven; raked piles shall be identified in the item descriptions and their inclination ratios stated (rule A2). Items for piles are deemed to include disposal of excavated material (rule C1).

The pile materials and section characteristics shall be given in item descriptions; section characteristics being the diameter for cast in place concrete piles, the cross-sectional area for preformed concrete or timber piles, and the mass per metre and cross-sectional dimensions for isolated steel piles.

With interlocking steel piles, the number of piles is not stated. Both the driving and materials items are measured by area, found by

166

multiplying the mean undeveloped horizontal lengths of pile walls by the depths (rule M7). The additional cost of corner, junction, closure and taper piles is covered in linear items (rule D7).

Rule M1 establishes the 'Commencing Surface' from which piles are driven or bored, as the surface adopted in the Bill of Quantities at which boring or driving is expected to begin. The measurement of cutting off surplus lengths or adding extensions is covered in Class Q.

Piling Ancillaries (Class Q)

Work incidental to piling operations, other than backfilling empty bores for cast in place concrete piles, is only measurable when expressly required (rule M1). Hence work undertaken by the Contractor at his own choosing will not be reimbursed unless it is covered in rates for items outside Class Q. Items for piling ancillaries are also deemed to include disposal of surplus materials unless otherwise stated (rule C1).

Work ancillary to piling is classified by pile type and size in a similar manner to Class P, for identification purposes, but ranges are used instead of actual cross-sectional dimensions. Cast in place piles may be concreted through a tremie pipe where water stands in the shaft, but this does not require specific mention.

Where the base of a cast in place pile is to be enlarged, the diameter shall be stated (rule A1), although the price will only cover the cost of the extra material around the shaft previously measured under Class P.

Pile extensions are measured in two items:

(1) number of pile extensions, to cover the cost of preparing piles to receive extensions and of making joints
(2) length of pile extensions to cover the cost of material in them, subdivided between those which do not exceed 3 m and those exceeding 3 m, and including the material to be used in the item description (rule A6).

The driving of extended piles is covered by Class P. Preparing heads of piles to receive permanent work is enumerated, while cutting off surplus lengths is taken as a linear item.

Removing obstructions (Q7) is priced on an hourly basis for breaking out rock or artificial hard material above the founding stratum of bored piles (rule M11). It is measured only when expressly required, when the Engineer can maintain control and keep records. Extraction of piles is classed as non-standard work, since it is not listed in Class Q, and suitable codes will be Q 3–6 9 *.

Note the pile testing items in Q8 differentiating between loading tests and non-destructive testing.

Worked Examples

Worked examples follow covering the measurement of concrete and timber piles and steel sheet piling.

For layout reasons, and for ease of reading, this page has intentionally been left blank.

DRAWING NO. 12

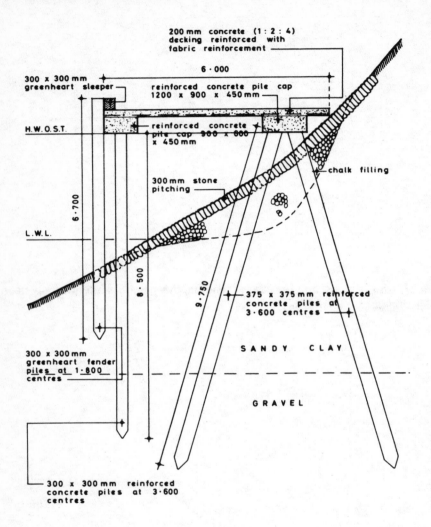

200 mm concrete (1 : 2 : 4) decking reinforced with fabric reinforcement

6 · 000

300 x 300 mm greenheart sleeper

reinforced concrete pile cap 1200 x 900 x 450 mm

H.W.O.S.T.

reinforced concrete pile cap 900 x 600 x 450 mm

chalk filling

300 mm stone pitching

6 · 700

L.W.L.

8 · 500

9 · 750

375 x 375 mm reinforced concrete piles at 3 · 600 centres

SANDY   CLAY

300 x 300 mm greenheart fender piles at 1 · 800 centres

GRAVEL

300 x 300 mm reinforced concrete piles at 3 · 600 centres

SECTION   THROUGH   QUAY

CONCRETE   AND   TIMBER   PILING

SCALE   1 : 100

170

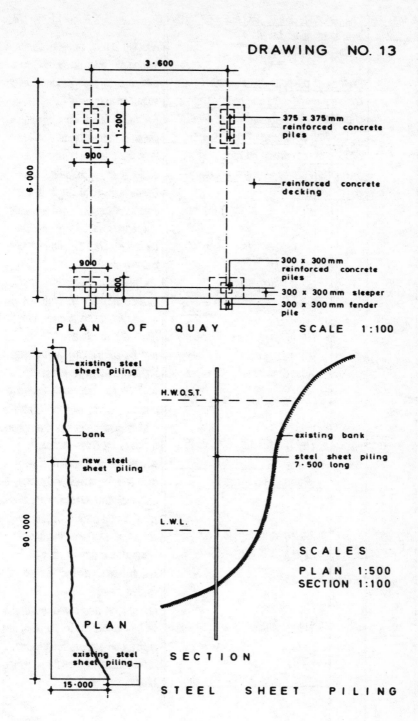

3·600

1·200

375 x 375 mm
reinforced concrete
piles

900

6·000

+ reinforced concrete
decking

900

300 x 300 mm
reinforced concrete
piles

600

300 mm sleeper

300 x 300 mm fender
pile

PLAN   OF   QUAY   SCALE 1:100

existing steel
sheet piling

H.W.O.S.T.

bank

existing bank

new steel
sheet piling

steel sheet piling
7·500 long

90·000

L.W.L.

SCALES

PLAN 1:500
SECTION 1:100

PLAN

existing steel
sheet piling

SECTION

15·000

STEEL   SHEET   PILING

171

(108 m length of quay taken)

PILES

Preformed piles, as quay support, concrete grade C25 as Specification clause 28·2; reinforcement as detail 3, Drawing J8/3, csa 0·1 - 0·15 m², 375 x 375 mm driven on rake at an inclination of 75° in pairs w. Commg. Surf. HWOST.

3·600) 108·00
30+1

	exposed len.	9·750
	add for strippg. end & connectg. to pile cap & deck slab.	600
		10·350

31/   2   Nr. of piles, len. 10·35m inc. m.s. drivg.   P341 hds. & shoes.

8·500
9·250
2)17·750
8·875

31/2/   8·88   Depth driven.   P342

Piling Ancillaries

31/2/   0·30   Cuttg. off surplus lens; 375 x 375 mm.
    Q374

31/   2   Prepg. heads; 375 x 375 mm.
    Q384

Adopt a logical sequence in the 'taking off', such as concrete piles and caps, concrete decking, timber piles and work to the embankment. Assuming this is an independent length of quay, allowance will have to be made for the additional work at the far end of the quay.

Dividing the total length by the spacing of the piles gives the number of spacings, as distinct from the number of piles (31 pairs of 375 x 375 mm piles).

The pile item descriptions shall include the materials (rule A1), inclination ratios of any raking piles (rule A2), structure to be supported and the Commencing Surface (rule A3). Preformed concrete pile prices include reinforcement, chamfered corners, tapered toes, moulds, shoes and related items. The description includes the cross-sectional dimensions or diameters of piles (rule A6), regardless of the cross-sectional area and diameter ranges given in the second division.

The cost of provision, use and removal of a driving rig will probably be covered by method-related charges.

Follow with any piling ancillaries; in this example a linear item for cutting off surplus lengths and an enumerated item for preparing heads.

10·1

		Preformed piles, as quay support, concrete grade C25 as Specification clause 28·2; reinforcement as detail 4, Drawing J8/3, csa 0·05 - 0·1 m², 300 x 300 mm w. Commg. Surf. HWOST.	Start with a suitable heading encompassing all the particulars listed in class P, including the cross-sectional area range and the cross-sectional dimensions.

$$\begin{array}{r} 8 \cdot 500 \\ 600 \\ \hline 9 \cdot 100 \end{array}$$

31/	1	Nr. of piles, len. 9·1 m inc. m.s. drivg. hds. & shoes. P331	The 300 x 300 mm piles fall into a different cross-sectional area range than the 375 x 375 mm piles (the respective cross sectional areas being 0·141 and 0·09 m²).
31/	5·10	Depth driven. P332	

Pilg. Ancillaries

31/	0·30	Cuttg. off surplus lens; 300 x 300 mm. Q373	Follow with piling ancillaries as with the 375 x 375 mm piles.
31/	1	Prepg. heads; 300 x 300 mm. Q383	

IN SITU CONCRETE
Pile caps

31/	1·20 0·90 0·45	Provsn. of conc; designed mix, grade C15, ct. to BS12, 20 mm agg. to BS 882. F123	Provision and placing of concrete are separated.
31/	0·90 0·60 0·45	& Placg. of conc. reinfd. pile caps; thickness: 300 - 500mm. F623	Pile caps are included in the same second division classification as bases and ground slabs, and the appropriate thickness range must be included in the item description.

10.2

		Description	Commentary
		Conc. ancillaries Fwk. fair fin; pile caps	Wrought formwork has been measured to the pile caps as they will be exposed.
31/	1·20 0·90	Hor. width : 0·4 – 1·22 m. (soffs. G214	No adjustment has been made for heads of piles or the cutting on formwork around them.
31/	0·90 0·60		Note build up of length of enclosing formwork in waste.

$$2\left/\begin{array}{l}1\cdot200\\900\\ \hline 2\cdot100 \\ \hline 4\cdot200\end{array}\right. \quad 2\left/\begin{array}{l}900\\600\\ \hline 1\cdot500 \\ \hline 3\cdot000\end{array}\right.$$

		Description	Commentary
31/	4·20 0·45	Vert. width : 0·4 – 1·22 m. (sides	The side forms are measured in m² as they exceed 200 mm wide.
31/	3·00 0·45	G244	
		Reinforcement ms bars to BS4449	4 nr. 25 mm reinforcing bars have been taken in each direction to each pile cap. Reinforcing bars are classified in the diameters listed in the third division and measured in tonnes.

$$\begin{array}{r}\text{less}\\\text{cover}\,2/40\end{array}\ \begin{array}{rrr}1\cdot200 & 900 & 600\\ 80 & 80 & 80\\ \hline 1\cdot120 & 820 & 520\end{array}$$

		Description	
31/4/	1·12	Diam. 25 mm.	
31/4/	0·82		
31/4/	0·82		
31/4/	0·52	G517	
		R.c. decking	
	108·00 6·00 0·20	Provsn. of conc; designed mix grade C15, ct. to BS12, 20 mm agg. to BS882. & F233	Follow with reinforced concrete decking, adopting the same sequence as for the pile caps.
		Placg. of conc., reinfd. susp. slab; thickness : 150–300 mm. F632	The placing item distinguishes between mass and reinforced concrete and the relevant thickness range must be stated.

$$\begin{array}{r}\text{less}\\\text{cover}\,2/20\end{array}\ \begin{array}{rr}108\cdot00 & 6\cdot00\\ 0\cdot40 & 0\cdot40\\ \hline 107\cdot60 & 5\cdot60\end{array}$$

		Description	Commentary
		Reinft.	
	107·60 5·60	Fabric high yield reinft. to BS4483, nom. mass 2–3 kg/m², ref. A142. G562	Fabric reinforcement is measured in m², giving the fabric reference as rule A9 of class G.

10.3

	108·00	Conc. ancillaries		
	6·00	Fwk. fair fin.		
		Hor.	G215	
31/	1·20	Ddt. ditto.		
	0·90			
31/	0·90		(pile caps	
	0·60		G215	
2/	108·00	Vert. width : 0·2 m. (edges.		
2/	6·00		G242	

As the width of the decking exceeds 1·22 m it is unnecessary to state a width of formwork. Deductions are made for the area of pile caps as each exceeds 0·5 m² (rules M4 & D3).

The formwork to the edges of the decking is measured as a linear item.

Formwork to temporary surfaces, such as construction joints, at the discretion of the Contractor, is not measured separately (rule M2c).

TIMBER
Hwd. components, csa
0·04 – 0·1 m²; wrought fin. & chfd. upper edges.

	108·00	300 x 300 mm greenheart, len: 3–5 m; sleeper. 0143

$$1·200 \overline{)108·000}$$
$$90+1$$

Fittgs. & fastengs. galvd. mild steel.

91/	1	Bolts; len: 500 mm, diam. 25 mm. 0540

&

Plates; 50 x 50 mm, thickness: 6 mm. 0550

It is necessary to state the nominal gross cross-sectional dimensions, grade or species of timber and any impregnation requirements or special surface finish in the item description (rule A1 of class O).

Separate items are not required for fixing timber components, or for boring, cutting and jointing (rule C1 of class O).

The sleeper is fixed at 1·2 m centres. The materials, types and sizes of fittings and fastenings shall be stated in item descriptions (rule A4 of class O).

PILES   Fender piles
add. len. to be ringed   6·700
& removed after driving   600
$$\overline{7·300}$$

$$1·800 \overline{)108·000}$$
$$60+1$$

Calculation of length and number of timber piles in waste (two fender piles in each 3·6 m length of quay).

10.4

		Timber fender piles csa 0·05–0·1m²; 300 x 300 mm greenheart, w. Commg. Surf. 1 m above HWOST.	The measurement of timber piles follows the same procedure as for preformed concrete piles.
61/	1	Nr. of piles, len: 7·3 m; gms drivg. hd. & shoe as detail D drg. FQ/17/C. P631	Details of driving heads and of shoes shall be stated in item descriptions for the number of
61/	2·00	Depth driven. P632	piles (rule A8 of class P).
		Pilg. ancillaries	Follow with piling ancillaries.
61/	0·60	Cuttg. off surplus lens; 300 x 300 mm. Q473	
61/	1	Preparg. heads; 300 x 300 mm. Q483	
61/	1	Bolts gms; len: 600 mm, diam. 25 mm. O540 &	The bolts are enumerated giving relevant particulars as rule A4 of class O. This is assuming no great variation in the cross-sectional
		Plates gms; 50 x 50 mm, thickness: 6mm. O550	area of the filling. In practice the volume would normally be computed from a number of cross
		EARTHWORKS Work to Embankment mainly below hwl.	sections. The triangular cross-sectional area is obtained from 'give and
½/	108·00 5·70 1·35	Fillg. to embankment, imported rock. (chalk fillg. E627	take' lines drawn across the boundaries.
	108·00 9·30	Pitching, imported rock; thickness: 300 mm, at L of 10° to 45° to hor. E647	Both the thickness and the slope of the pitching are to be stated (rules A13 and A14 of class E).

10.5

		(permanent)

Interlockg. steel piles grade —,
type —, section modulus
1150 cm³/m, Commg. Surf. 900mm
above HWOST.

The section reference or mass
per metre and section modulus
are required for interlocking
steel piles (rule A10 of class P).
The Commencing Surface
should also be stated (rule M1).

| 2/ | 7·50 | Len. of junctn. piles;
connectn. to xtg.   P831·1 |
|---|---|---|

| | 7·50 | Len. of corner piles.
P831·2 |
|---|---|---|

90·000
15·000
105·000

Corner, junction, closure and
taper piles are classified as
special piles and measured as
linear items (rule D7).
It is necessary to add suffixes
to the codes in these cases.

| | 105·00
1·50 | Driven area.   P832 |
|---|---|---|

The driven area is measured
separately from the total area
of piling computed in
accordance with rule M7.

| | 105·00
7·50 | Area of piles of len: n.e.
14m; treated w. 2 cts. of
bitumen paint.   P833 |
|---|---|---|

<u>Obstructns.</u>

| | 10 h | Breakg. out rock.
Q700 |
|---|---|---|

This work is measured by
time to provide an operative
rate.

11.1

10 Measurement of Timber and Associated Work

Timber (Class O)

The rules for measurement of work in this class cover timber components, timber decking, and metal fittings and fastenings to the timberwork. The timber components are those used in permanent civil engineering work such as jetty timbers and fendering. Carpentry and joinery work to buildings is measured in accordance with Class Z.

The approach to the measurement of timberwork is straightforward with decking measured in m² (void allowance of 0.5 m²) and components by length (m). The nominal gross cross-sectional dimensions or thicknesses (unplaned), grade or species, impregnation requirements or special surface finishes shall be stated in item descriptions (rule A1). In addition, the structural use and location of timber components exceeding 3 m in length are to be stated in item descriptions (rule A2).

Metal fittings and fastenings are enumerated under the categories listed in the second division, and the materials, types and sizes of fittings and fastenings shall be stated in item descriptions (rule A4). Separate items are not required for fixing timber components and decking, or for boring, cutting and jointing (rule C1).

Miscellaneous Work (Class X)

This class is concerned with the measurement of fences, gates and stiles, and drainage to structures above ground, such as gutters and downpipes, and rock filled gabions. Gabions are wire or plastic mesh cages filled with loose rock or crushed stone and common sizes are 2 × 1 × 1 m and 2 × 1 × 0.5 m. They are used extensively for revetment and linings in sea and river defences.

Item descriptions for fences and associated work shall give their type, and principal dimensions, and also of foundations where appropriate (rules A2 and A3 of Class X). Fences are measured as linear items and gates and stiles are enumerated. The height and width classification ranges are for coding purposes only, and actual heights and widths will be given in item descriptions. Fences erected to a curve of a radius not exceeding 100 m or on a surface inclined at an angle exceeding 10° require specific mention because of the additional cost (rule A1). Items for fences are deemed to include excavation, preparation of surfaces, disposal of excavated material, upholding sides of excavation, backfilling, removal of existing services, concrete, formwork and reinforcement (rule C1) and also end posts, angle posts, straining posts and gate posts (rule C2).

Gutters and downpipes are measured as linear items, including holderbats and brackets, but fittings such as bends, angles, stopends, outlets, swan necks and shoes are enumerated (rule D3 of Class X). Item descriptions shall include the type, materials and principal dimensions of components (rule A5 of Class X).

Item descriptions for rock filled gabions shall include the particulars listed in rule A6 of Class X. Rock filled gabions exceeding 300 mm thick are classed as boxed gabions and those not exceeding 300 mm thick as mattress gabions.

Worked Example

A worked example follows covering the measurement of a timber jetty.

TIMBER JETTY

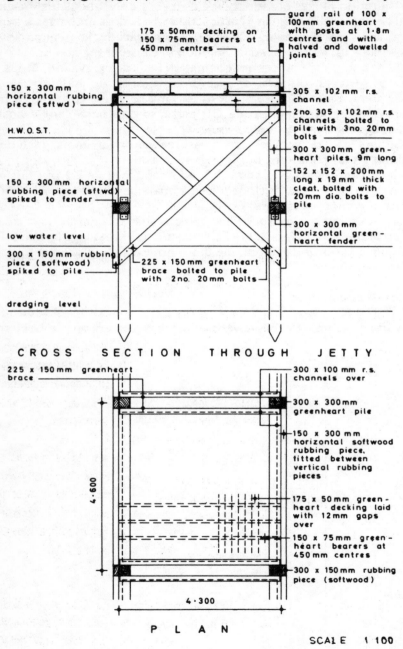

guard rail of 100 x 100mm greenheart with posts at 1·8m centres and with halved and dowelled joints

175 x 50mm decking on 150 x 75mm bearers at 450mm centres

150 x 300mm horizontal rubbing piece (sftwd)

305 x 102mm r.s. channel

2no. 305 x 102mm r.s. channels bolted to pile with 3no. 20mm bolts

H.W.O.S.T.

300 x 300mm green-heart piles, 9m long

150 x 300mm horizontal rubbing piece (sftwd) spiked to fender

152 x 152 x 200mm long x 19mm thick cleat, bolted with 20mm dia. bolts to pile

low water level

300 x 300mm horizontal green-heart fender

300 x 150mm rubbing piece (softwood) spiked to pile

225 x 150mm greenheart brace bolted to pile with 2no. 20mm bolts

dredging level

CROSS SECTION THROUGH JETTY

225 x 150mm greenheart brace

300 x 100mm r.s. channels over

300 x 300mm greenheart pile

150 x 300mm horizontal softwood rubbing piece, fitted between vertical rubbing pieces

4·600

175 x 50mm green-heart decking laid with 12mm gaps over

150 x 75mm green-heart bearers at 450mm centres

300 x 150mm rubbing piece (softwood)

4·300

PLAN

SCALE 1 100

180

		TIMBER JETTY (92 m length)	EXAMPLE XII

PILES

Timber piles supportg. jetty, csa 0·05 − 0·1 m²; 300 × 300 mm greenheart, w. Commg. Surf. 1·20 m above HWOST.

The cross-section type and cross-section dimensions shall be stated in the item description (rule A6 of class P).

$$4·600 \overline{)92·000}$$
$$20+1$$

add len. for ringing & removal after driving.
$$9·000$$
$$\underline{600}$$
$$9·600$$

Add 1 to allow for a pair of piles at each end of the jetty.

21/   **2**

Nr. of piles, len: 9·6 m, gms drivg. hd. & shoe as detail C drg. RJ/5/1.   P631

Two separate items are required for each group of timber piles.
(1) number of piles of stated length
(2) depth driven.

21/2/   **3·00**

Depth driven.   P632

The lengths of timber piles include heads and shoes (rule D3 of class P).

Pilg. ancillaries

21/   **0·60**

Cuttg. off surplus lens.; 300 × 300 mm.   Q473

Follow with the work to the tops of piles.
No separate items are required for pointed ends or ringing heads. Information with regard to water level will be included in the preamble.

21/   **2**

Preparg. heads; 300 × 300 mm.   Q488

TIMBER

Hardwood components, csa 0·04 − 0·1 m²; wrought finish.

Descriptions of timber components shall state gross cross-sectional dimensions, grade or species, any impregnation requirements or special surface finishes (rule A1 of class O).

fenders
$$4·600$$
less piles²/150 $$\underline{300}$$
$$4·300$$

20/2/   **4·30**

300 × 300 mm greenheart, len: 3−5 m; hor. fender.   O143

The description includes gross (unplaned) dimensions, species of timber, length range and function.

12.1

TIMBER JETTY (Contd.)

		Hardwood components, csa 0.02 - 0.04 m²; wrought finish.
21/2/	6.05	225 x 150 mm greenheart, len: 5 - 8 m; braces. 0134
		Fittgs. & Fastengs. galvd. mild steel.
20/2/4/	1	Cleats ; 152 x 152 x 19 mm L, len: 200 mm. (fenders/piles 0590·1
		fender 300
		cleats 38
		clearance 40
		378
20/2/4/	1	Bolts ; len: 380mm, (fenders diam: 20 mm. 0540·1
21/2/4/	1	Bolts ab. (piles 0540·1
		&
		Plates: 50 x 50 mm, thickness 6 mm. 0550·1 pile 300
		brace 150
		clearance 40
		490
21/2/4/	1	Bolts: len: 500mm; diam: 20mm. 0540·2
		& (braces/piles
		Plates ; 50 x 50 mm, thickness 6 mm. 0550·1
		Softwood components; csa 0.04 - 0.1 m², wrought finish, pressure creosoted.
2/	92·00	150 x 300 mm len: 3 - 5 m; hor. rubbg. piece w. rdd. / top hor. edge. 0243·1 (rubbg pcs
20/2/2/	2	Spikes gms, len: 225 mm. 0520·1

The cross-sectional area of each brace is 0.034 m² and hence it falls in the 0.02 - 0.04 m² range. The splayed ends are included in the timber rates (rule C1 of class O).

Cleats would come within the fittings and fastenings classification but there is no specific mention of them in the second division; hence the use of the digit 9.

Materials, types and sizes of fittings and fastenings shall be stated in item descriptions (rule A4 of class O).

Plates are measured separately from the bolts as prescribed in the second division.

21 pairs of braces with 4 bolts to each brace.

The cross-sectional area is 0.045 m² and it thus comes within the range 0.04 - 0.1 m². The location of the component and any labours should be included in the item description.

Additional locational notes are often inserted in waste to aid identification.

Horizontal rubbing piece is spiked to heads of piles (2 spikes at each end of each piece).

12.2

TIMBER JETTY (Contd.)

20/2/	4.30	150 x 300 mm, len: (lower hor. rubbg. piece) 3-5m; hor. rubbg. piece.	0243.2
20/2/2/	2	Spikes, gms, len: 225 mm.	0520.1
21/2/	4.50	300 x 150 mm, len: 3-5m; vert. rubbg. piece.	0243.3
21/2/	8	Spikes, gms, len: 225 mm.	0520.1

STRUCTURAL METALWORK
Steel grade 43A

21/2/	4.30	Fabrication of members for frames, beams, st. on plan; channel 305 x 102 mm X 46.14 kg (deckg. brrs.)
		& M321
		Erection of members for frs, perm. erectn, ditto. M620

	len.
pile	300
2 channels	20
clearance	40
	360

21/2/3/	1	Site bolts: black; len: 360 mm, diam: 20 mm. M632.1
4/	92.00	Fabricatn. of members for frames, beams, st. on plan; channel 305 x 102 mm X 46.14 kg. M321
		&
		Erection of members for frs. perm. erectn, ditto. M620

The lower horizontal rubbing pieces run between the vertical rubbing pieces (20 lengths on each side).

Spikes are enumerated, stating the material and size (rule A4 of class O).

There are 21 pairs of piles and vertical rubbing pieces.

The vertical rubbing pieces are spiked to the piles. Alternatively, all the spikes could be taken together under a heading of fittings and fastenings.

The steel members are taken as linear items, stating the weight in kg/m. They are subsequently weighted up and entered in the Bill in tonnes as fabrication and erection items. Drilling holes are not enumerated as they are included in the steelwork rates.

Bolts through webs on 2 channels and head of pile.

Items for site bolts are deemed to include supply and delivery to site (rule C3 of class M).

Channels running the full length of the jetty.

Proceed in a logical sequence with the fabrication and erection of the structural metalwork, using the descriptions prescribed in class M.

12.3

TIMBER JETTY (Contd.)

				Len.
			2 channel flanges	30
			clearance	40
				70

| 21/4/2/ | 1 | | Site bolts: black; len. 70 mm, diam: 20 mm. | One bolt taken at each connection. |

M632·2

TIMBER

Hardwood components csa 0·01–0·02 m². unwrought finish.

			Deckg.
		450) 92·000	
		205+1	

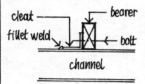

cleat — | bearer
fillet weld | bolt
channel

| 206/ | 4·30 | | 150 x 75 mm greenheart, len: 3 – 5 m; deckg. brrs.   0123 | Bearers are fixed to channels with steel angle cleats welded to channels and bolted to bearers. |

Fittgs. & Fastengs. galvd. mild steel.

| 206/4/ | 1 | | Cleats; 76·2 x 76·2 x 9·4 mm L, len: 100 mm.   0590·2 | Four connections to each bearer. Welding of cleats will most probably be carried out at the fabricator's works. Build up of length of bolt in waste. |

			bearer	75
			angle	10
			clearance	40
				125

| 206/4/ | 2 | | Bolts; len: 125mm, diam: 12mm.   0540·3 | Two bolts to each connection. Third different size of bolt and hence addition of suffix 3 to code reference. |

width of deck   4·300
total width of gaps
= 1/15 x 4·300   =   287
(allowance for 12·5 mm
gap between 175 mm decking
members is 1:15 gap)

                          4·300
             less gaps   287
                          4·013

The effective width of hardwood decking is obtained by deducting the gaps between the decking members, as each gap exceeds 0·5 m² in area (rule M2 of class O).

12.4

TIMBER JETTY (Contd.)

Hardwood decking; thickness:
50 mm; wrought finish.

92·00	Greenheart 175mm widths	The description includes the
4·01	septd. by 12·5mm gaps	timber species and surface finish.

Greenheart 175mm widths
septd. by 12·5mm gaps
(mesd. net).   0320

Hardwood components; csa n.e. 0·01m²;
wrought finish.

Guardrail

1·800) 92·000
        51+1

2/3/	92·00	100 x 100 mm greenheart
2/3/52/	0·10	len: 1·5 – 3 m; guardrails.

100 x 100 mm greenheart
len: 1·5 – 3 m; guardrails.
(laps
0112

| 2/52/ | 1·30 | 100 x 100 mm greenheart, |

100 x 100 mm greenheart,
len: n.e 1·5m; posts to
guardrail.   0111

Fittings & Fastenings
galvd. mild steel.

| 52/2/ | 1 | Cleats; 76 x 76 x 9 mm |

Cleats; 76 x 76 x 9 mm
L, len: 75mm.   0590·3

post   100
angle   10
clearance   40
            150

| 52/2/ | 1 | Bolts; len: 150 mm, (to posts |

Bolts; len: 150 mm, (to posts
diam: 12 mm.   0540·4

bearer   150
angle   10
clearance   40
           200

| 52/2/ | 1 | Bolts; len: 200 mm, (to brrs. |

Bolts; len: 200 mm, (to brrs.
diam: 12mm.   0540·5

12.5

The description includes the
timber species and surface finish.
The actual thickness has been
given instead of the range as
only one thickness is involved.
The gaps need mentioning as they
will increase the fixing costs.
There are three rails on each side
of the quay. The number of posts
is calculated in waste.
The guardrail is broken down into
component parts and measured
as linear items. The timber rates
are deemed to include forming
joints (rule C1 of class O).

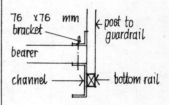

Double fixing of guardrail at
1·80 m centres.
(1) bottom rail bolted to channel
(2) post bolted to bracket which
    is bolted to 150 x 75 mm
    bearers.

The dimensions of the angle
cleats have now been changed.
The person measuring the work
should always be on the look-
out for discrepancies of this
kind.

TIMBER JETTY (Contd.)

		bottom rail 100
		channel web 10
		clearance 40
		————
		150

52/2/	1	Bolts; len: 150mm,	A similar size of bolt to that
		diam: 12 mm. (bolt rail	measured earlier and so it
		(to chann.	carries the same code reference.
		0540·4	

PAINTING

The measurement of painting
of steelwork follows.
It is assumed that the steelwork

       305
2/102   204
    2/ 509
      1·018

4/	92·00	② bit. paint, met. sectns.	is to be painted by the main
	1·02		contractor after erection by him,
			and is therefore measured in
21/2/	4·30		accordance with Class V.
	1·02	(channs.	Girth of channel = twice height
		V870	+4 times width.

Item descriptions state the
materials used and either the
number of coats or the film
thickness (rule A1 of class V).
The painting of metal sections
is deemed to include painting
the surfaces of connecting plates,
brackets, rivets, bolts, nuts and
similar projections (rule C2 of
class V). Hence separate
painting items are not required
for cleats. Preparation of
surfaces before painting is
deemed to be included unless
there is more than one type of
preparation, when they shall
be described in the item
description (rules C1 and A2 of
class V).

12.6

11 Measurement of Metalwork

Structural Metalwork (Class M)

A considerable proportion of the cost of steelwork is in fabrication where activities such as welding on fillets and cutting holes are entailed. Hence detailed drawings of connections and fittings are necessary at the tendering stage and this is assumed in the rules for measurement.

Barnes[5] has shown just how much detail fabricators require for tendering purposes, since although steel is purchased by the tonne, the price will vary according to the section, size, length, quantity, quality, finish, and the requirements for testing and inspection. The rate per tonne for plates will also be influenced by their length, width and thickness.

Shop and site bolts are expensive and their number requires accurate assessment. Drawing office costs vary with the number of separate pieces of steel, their complexity and the number of drawings required. Fabrication costs are influenced by a whole range of production processes.

Erection costs are affected by many matters, including the number, size and weight or mass of pieces of steel and their location and the form of connections, in addition to site conditions and facilities.

The drawings from which the quantities are taken should be inserted at the heads of sheets and the quantities listed under the division and description headings in Class M. A brief description should be given of each class of member, including the identification marks on the drawing and grade of metal. Particulars of each component should include the number required, length, type of section, size of section and unit mass or thickness, to compute the mass of individual components and fittings and finally the total mass of the items. Surface areas are required for the surface treatments listed in M8 1–7.

Fittings such as caps, bases, gussets, end plates, cleats, brackets, stiffeners, distance pieces, separators and packings are listed and weighted with the main member or unit to which they are attached. The second division classifications of trusses, purlins and cladding rails

187

can be applied to engineering structures such as conveyor gantries, and, where they form part of a building, the building method of measurement may be considered more appropriate.

CESMM3 distinguishes between fabrication and erection, with sub-division of fabrication items into a number of broad categories, such as columns, beams, trestles, towers, bracings and grillages. In both categories of work there are three broad divisions: members for bridges, members for frames, and other members (usually isolated ones). Item descriptions for fabrication shall identify tapered and castellated members, (rule A2).

The mass of members, other than plates or flats, is calculated from their overall lengths, with no deductions for splay cut or mitred ends (rule M2). No allowance is made for rolling margin, or the mass of weld fillets, bolts, nuts, washers and rivets (rule M4), or voids less than 0.1 m^2 (rule M5), and all fillets and connections are included in the metal-work rates.

Trestles, towers, built-up columns, trusses and built-up girders can be made from sections and/or plates and may be of compounded sections, lattice girders, plate girders or box type construction. Details of the members shall be given in accordance with rule A4. Light crane rails are generally included with the main beams or girders to which they are attached, while rails for heavier cranes (over 20 t capacity) are best kept separate together with their ancillary fittings, such as fixing clips and resilient pads. Anchorages and holding down bolt assemblies are suitably described and enumerated (rule M7).

Off-site surface treatment of metalwork is measured in m^2 under the classifications listed in the second division. Where blast cleaning is specified, the standard of finish should be stated, for example second quality to BS 4232. Painting systems shall also be clearly defined. Surface treatments carried out on site after the erection of structural metalwork are measured in accordance with the rules in Class V (Painting) (rule M8).

Testing will be covered under General Items and the supply, delivery, unloading, operation and dismantling of cranes and plant can be covered by erection rates or be included in method-related charges.

Miscellaneous Metalwork (Class N)

This class covers metal components not specifically included elsewhere in CESMM3. Separate items are not given for erection and fixing or for the provision of fixings (rule C1). Item descriptions shall include the specification and thickness of metal, surface treatments and the principal dimensions of miscellaneous metalwork assemblies (rule A1). Alternatively, a more effective approach is often to identify the work by reference to material, construction and assembly details given in drawings and/or specification. This latter approach is particularly well suited to stairways and walkways, to avoid lengthy bill descriptions, using mark numbers for identification purposes as described in the footnote on page 63 of CESMM3.

There is a variety of units of measurement ranging from stairways and walkways in t; handrails, bridge parapets, ladders, walings and frames (measured on external perimeters) in m; cladding, flooring, panelling and duct covers in m²; and tie rods, bridge bearings and tanks by nr. The schedule of items listed in this class is not comprehensive and item codes N3–8 are available for the inclusion of non-standard components. Most of these are likely to be measured in tonnes identified by drawing references. No deductions are made for voids not exceeding 0.5 m² (rule M3). Sophisticated cladding is more likely to be measured in accordance with the building method.

Worked Example

A worked example follows covering the measurement of a steel-framed gantry.

STEEL – FRAMED   GANTRY

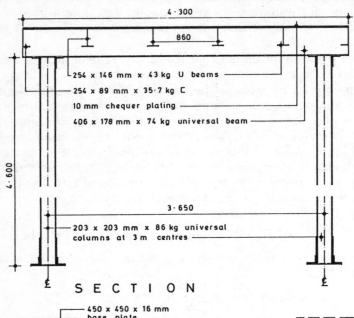

4 · 300

860

254 x 146 mm x 43 kg U beams

254 x 89 mm x 35·7 kg C

10 mm chequer plating

406 x 178 mm x 74 kg universal beam

4 · 600

3 · 650

203 x 203 mm x 86 kg universal
columns at 3 m centres

S E C T I O N

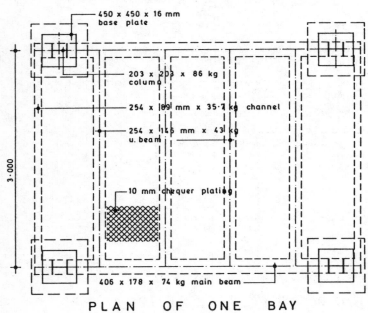

450 x 450 x 16 mm
base plate

203 x 203 x 86 kg
column

254 x 89 mm x 35·7 kg channel

254 x 146 mm x 43 kg
u. beam

3 · 000

10 mm chequer plating

406 x 178 x 74 kg main beam

P L A N   O F   O N E   B A Y

SCALE   1 : 50

190

STEEL - FRAMED GANTRY
DRAWING NO. 16

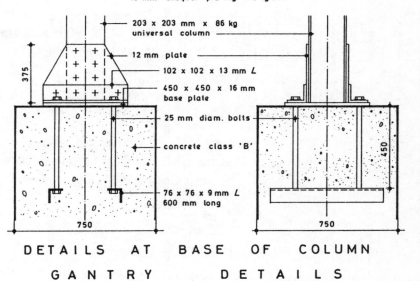

254 x 89 mm x 35·7 kg
r.s. channel

10 mm chequer plating

102 x 102 x 9 mm L
150 mm long

254 x 146 x 43 kg
universal beam

102 x 102 x 9 mm L
89 mm long

102 x 102 x 9 mm L
146 mm long

152 x 102 x 13 mm L
203 mm long

406 x 178 mm x 74 kg universal main beam

DETAIL AT HEAD OF COLUMN

WEIGHTS OF SECTIONS

152 x 102 x 13 mm L = 23·99 kg/m
102 x 102 x 13 mm L = 18·91 kg/m
102 x 102 x 9 mm L = 14·44 kg/m
76 x 76 x 9 mm L = 10·57 kg/m
10 mm chequer plating = 70 kg/m^2

203 x 203 mm x 86 kg
universal column

12 mm plate

102 x 102 x 13 mm L

450 x 450 x 16 mm
base plate

25 mm diam. bolts

concrete class 'B'

76 x 76 x 9 mm L
600 mm long

375

450

750

750

DETAILS AT BASE OF COLUMN
GANTRY DETAILS

SCALE 1:20

		(30 m length)

Fabrication of members for frames

CESMM3 subdivides the measurement of structural metalwork in frames into fabrication and permanent erection. The fabrication in this example is subdivided into columns and beams and the connections are weighted up with the appropriate members; adopting a logical sequence in the taking-off.

<u>Columns, straight on plan</u>

3·000)30·000
10+1

¹¹/₂/	4·60	203 x 203 mm X 86 kg U.C. M311

All members are taken off in metres to be subsequently reduced to tonnes, prior to billing.

At top of column (2 brackets to each column).

¹¹/₂/₂/	0·20	152 x 102 x 13 mm X 23·99 kg L. (stan. conns. M311

Note method of building up dimensions of irregular area of plate in 'waste'.

375   450
Less angle 102   203
273   2)653
     327

¹¹/₂/₂/	0·45	12 mm plate. M311
	0·10	
¹¹/₂/₂/	0·33	(stan. bases
	0·27	

The mass of mild steel is based on 785 kg/m^2 per 100 mm thickness (7·85t/m^3)(rule M6).

¹¹/₂/₂/	0·45	102 x 102 x 13 mm X 18·91 kg L. (stan. bases M311

All angles and plates in connections to columns will be weighted up with the columns.

¹¹/₂/	0·45	16 mm plate. (stan. base plates
	0·45	M311

¹¹/₂/₂/	0·60	76 x 76 x 9 mm X 10·57 kg L. (stan.bases M311

Angles at lower ends of holding down bolts set in concrete.

13.1

			Description	Notes
			Fabrication of members for frames	No allowance is made for the mass of weld fillets, bolts, nuts, washers, rivets and protective coatings (rule M4).
			<u>Beams, straight on plan</u>	
11/	4.30		406 x 178 mm x 74 kg U.B. (main beams M321	
10/2/	3.00		254 x 89 mm x 35.7 kg Channel. (subsid. beams M321	10 bays with 2 lengths of channel to each bay.
10/4/	3.00		254 x 146 mm x 43 kg U.B. (subsid. beams M321	It is unnecessary to deduct the very small thickness of the web in the main beam when determining the length of the subsidiary beams.
10/5/2/2/	0.15		102 x 102 x 9 mm x 14.44 kg L. M321	Side cleats to ends of beams and channels (both sides of main beams).
10/4/2/	0.15			brackets to beams
10/2/2/	0.09		(beam (conns.	brackets to channels
			<u>Erection of members for frames, permanent erection</u>	A single item for all steel members and associated connections will be provided in tonnes for erection.
			<u>Take the weighted up total for fabrication as erection</u> M620	Site bolts are enumerated with the length stated in this example because of their considerable length.
11/2/	4		Site bolts: HSFG general grade, diameter: 25 mm; length: 525 mm. M644	The casting of the bolts and angles into the concrete bases will be measured as an enumerated item of inserts under class G. The item description for the plate flooring shall state the specification and thickness of metal used (rule A1 of class N).
	30.00 4.30		Chequer plate floorg; thickness: 10 mm, mass: 70 kg/m². N170	

			Site Painting	Site painting after erection is measured in m² in accordance with class V.
			Zinc phosphate primer on metal sections after erection & 3 coats oil paint on ditto. V170	
			V370	
			Columns	Girth of column = twice the depth + 4 times the width. Most of the area of connecting brackets will be included in the area of beams.
		203 x 203	2/203 406 4/203 812 1·218	
11/2/	4·60 1·22		(cols.	
		Primer		
11/2/2/	0·45 0·10		(angles	Alternatively, the surface area of universal columns and beams, and channels can be extracted from Steelwork Tables.
		&		
11/2/2/	0·33 0·27		(outside of plates	
		③	450 203 247	
11/2/	0·25 0·10		(part of insides of angles.	Location notes are inserted in waste to help in identifying the particular items.
11/2/	0·45 0·45		(base plates	
			V170 and V370	
			Beams	
		406 x 178	2/406 812 4/178 712 1·524	Same approach to calculation of girth of beams as for columns.
		254 x 89	2/254 508 4/89 356 864	
		254 x 146	2/254 508 4/146 584 1·092	Surface treatment of structural metalwork prior to erection is also measured in m². Item descriptions for painting shall state the material used and either the number of coats or the film thickness (rule A1 of class V).
11/	4·30 1·52	Primer	(main beams	
10/2/	3·00 0·86	&	(channs.	
10/4/	3·00 1·09	③	(sub-beams	
			V170 and V370	

13.3

12 Measurement of Roads and Pavings

Class R in CESMM3 prescribes rules for the measurement of sub-bases, bases and surfacings of roads, airport runways, light-duty pavements, footways, cycle tracks and other paved areas, together with the necessary kerbs, channels and edgings, traffic signs and surface markings. Landscaping, drainage, fences and gates, and gantries and similar structures supporting traffic signs are measured in accordance with the appropriate classes.

The various courses of road materials in sub-bases, bases and surfacings are each measured separately in m^2, describing the material and giving the depth of each course or slab and the spread rate of applied surface finishes. The third division thickness ranges are overridden by rule A1 requiring the actual depth to be stated. Work to surfaces inclined at an angle exceeding 10° to the horizontal is so described and measured separately (rule A2). No deductions shall be made for manhole covers and the like less than 1 m^2 in area (rule M1).

The details of the construction work draw heavily on 'Specification for Highway Works'[31] (Department of Transport). Thus sub-bases of granular material may be either DTp Specified type 1 or type 2. Type 1 can consist of crushed rock, crushed slag, crushed concrete or well-burnt non-plastic shale within a specified grading range, whereas type 2 also includes natural sands and gravels and there are variations in the grading range. In like manner concrete carriageway slabs may be of DTp specified paving quality jointed reinforced concrete (JBC). This is concrete of grade C40 complying with BS 5328, with a minimum cement content of 320 kg/m^3 of ordinary Portland cement (OPC) and with the average value of any four consecutive test results at 7 days having a strength of not less than 31 $N/mm.^2$

Tolerances in surface levels and finishes have considerable impact on plant and labour costs and item descriptions need to be extended to cover differing or special tolerance requirements, in accordance with paragraph 5.10 of CESMM3.

With concrete pavements, item descriptions for steel fabric reinforcement to BS 4483 shall include the fabric reference, while descriptions of

195

other fabric reinforcement shall state the material, sizes and nominal mass/m^2 (rule A4). The area of additional fabric in laps is not measured (rule M3).

Separate items are not required for formwork to slabs or joints in concrete pavements (rule C1). Construction joints are measured only when they are expressly required (rule M7) and the dimensions, spacing and nature of components to joints shall be given in item descriptions (rule A6).

Kerbs, channels and edgings are measured as linear items including concrete beds and backings, with the details given in item descriptions (rule A7). Excavation for kerbs, channels and edgings should strictly be measured separately in accordance with Class E, but it is often more convenient to include it with the relevant items in Class R and to provide an appropriate preamble statement in accordance with paragraph 5.4 of CESMM3. The different cross-sections of precast concrete kerbs with bullnosed, 45° splayed and half battered faces relate to the prescribed details in BS 7263. Kerbs, channels and edgings laid straight or to curves with a radius exceeding 12 m are grouped together.

Traffic signs are enumerated giving the details listed in rule A8. Road studs are also enumerated while line surface markings are measured as linear items and in the case of intermittent markings shall exclude the gaps (rule M9).

Worked Example

A worked example follows covering the measurement of an estate road.

For layout reasons, and for ease of reading, this page has intentionally been left blank.

DRAWING NO. 17 ESTATE ROAD

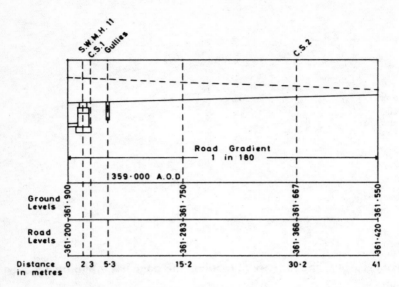

LONGITUDINAL SECTION

Road Gradient
1 in 180

359·000 A.O.D.

Ground Levels	361·900		361·750	361·667	361·550
Road Levels	361·200		361·283	361·366	361·420
Distance in metres	0 2 3 5·3		15·2	30·2	41

S.W.M.H. 11
C.S.1
Gullies
C.S.2

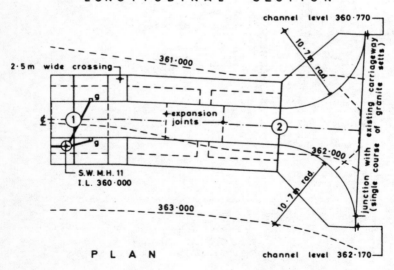

PLAN

2·5m wide crossing

361·000

expansion joints

S.W.M.H. 11
I.L. 360·000

363·000

channel level 360·770

10·7 m rad.

362·000

10·7 m rad.

channel level 362·170

junction with existing carriageway
(single course of granite setts)

SCALES : HORIZONTAL 1:500
VERTICAL 1:100

ESTATE ROAD DETAILS
DRAWING NO. 18

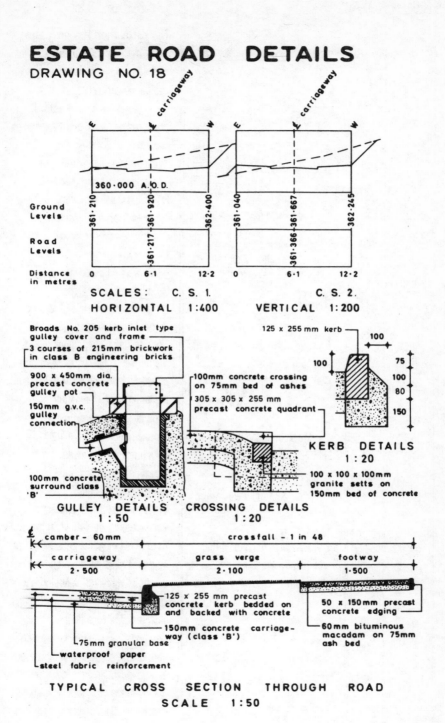

carriageway

E ℄ W   E ℄ W

360·000 A.O.D.

Ground Levels

361·210 361·920 362·400   361·040 361·667 362·245
361·217   361·366

Road Levels

Distance in metres

0   6·1   12·2   0   6·1   12·2

SCALES: C. S. 1.   C. S. 2.
HORIZONTAL 1:400   VERTICAL 1:200

Broads No. 205 kerb inlet type gulley cover and frame

3 courses of 215mm brickwork in class B engineering bricks

900 x 450mm dia. precast concrete gulley pot

150mm g.v.c. gulley connection

100mm concrete surround class 'B'

125 x 255 mm kerb

100

100   75
         100
         80
         150

KERB DETAILS
1 : 20

100mm concrete crossing on 75mm bed of ashes

305 x 305 x 255 mm precast concrete quadrant

100 x 100 x 100mm granite setts on 150mm bed of concrete

GULLEY DETAILS   CROSSING DETAILS
1 : 50   1 : 20

℄ camber – 60mm   crossfall – 1 in 48

carriageway   grass verge   footway

2·500   2·100   1·500

125 x 255 mm precast concrete kerb bedded on and backed with concrete

150mm concrete carriageway (class 'B')

75mm granular base

waterproof paper

steel fabric reinforcement

50 x 150mm precast concrete edging

60mm bituminous macadam on 75mm ash bed

TYPICAL CROSS SECTION THROUGH ROAD
SCALE 1:50

199

EXAMPLE XIV

Earthworks
Carriageway
av. depth. of excavn.

	CS1	CS2
	361·920	361·667
	361·217	361·366
add road thickness	703	301
	225	225
	928	526
		928
		2)1·454
av. depth of excavn.		727

width
carriageway 5·000
add kerbs 2/225 450
5·450

bellmouth depths
- 70 East side (fill
350 W. side
2)280
add road thickness 140
225
365

The excavation has been taken separately for carriageway, kerbs and footways, because of varying depths in each case.

The measurement rules are prescribed in Class E and the sequence adopted is as follows:

(1) excavation for cuttings for carriageway and disposal.
(2) adjustment for kerbs.
(3) excavation for cuttings for footways and disposal
(4) excavation for cuttings for banks and disposal
(5) adjustment for topsoil
(6) trimming, soiling and seeding banks and verges.

The additional area of one side of the bellmouth $= 3/14 \times \text{radius}^2$ (area of square with side equal to length of radius less area of quadrant or ¼ circle of same radius).

Topsoil will be adjusted later. Alternatively, the whole area of topsoil to be stripped could be measured at the outset.

2/3/14/	41·00	Excavn. for cuttgs; Commg.
	5·45	Surf. u/s of topsoil. E220
	0·73	
		& (bellmouth
	10·70	
	10·70	Disposal of excvtd. mat.
	0·37	E532

41·000
less rad. kerb 10·700
30·300

With excavation for cuttings, it is not necessary to state depth ranges. Excavated material is deemed to be material other than topsoil, rock or artificial material, unless otherwise described (rule D1).

Fill is a separate item.

The additional excavation for kerb foundations below road formation is kept separate from excavation for cuttings as it will be a more expensive item, probably involving hand excavation.

2/	30·30	Excavn. for foundations;
	0·23	max. depth: ne 0·25 m; Commg.
	0·08	Surf. road formatn. (kerbs
2/½/22/7/		E321·1
	10·70	& (bellmouth
	0·23	
	0·08	Disposal of excvtd. mat.
		E532

14.1

200

| | Footways | |
| --- | E. side | W. side |

av. depth at CS1   60   900
add thickness of path or verge   135
             195

av. depth at CS2 – 250 (fill   700
less thickness of path or verge   135
         – 115 (fill
          2)1·600
            800
add thickness of path or verge   135
            935

E. side
CS1   195
CS2 topsoil   150
       2)345
         172

            width
path   1·500
verge   2·100
         3·600
less kerb & backg.   225
         3·375

30·30	Excavn. for cuttgs; (W. side
3·38	Commg. Surf. u/s of
0·94	topsoil. E220

30·00	& (E. side
3·38	
0·17	Disposal of excvtd. mat.
	E532

Footways at bellmouth
	E. side	W. side
CS2	– 250	700
extremity	– 407	543
	2)– 657	2)1·243
	– 328	622

add thickness of path or verge   135   135
         – 193 (fill   757

The fill required under footways and verges on the east side will be made up of non-selected excavated material.

The whole of the area of paths and verges in normally stripped of topsoil, so that some excavation is required even in places which will subsequently receive fill.

Excavation for kerbs and backing has already been taken with the carriageway and so needs deducting from the overall width of path and verge.

All excavated material for disposal is taken as material other than topsoil, rock or other hard material in the first instance, and the necessary adjustments will be made later.

The depths to surface of paving at extreme ends of paths are calculated thus:

	E. side	W. side
channel lev.	360·770	362·170
add depth of kerb + ½ fall on path	137	137
	360·907	362·307

ground lev. at centre of path
(interpolated)	360·500	362·850
less finished level	360·907	362·307
depth	– ·407 (fill)	·543

14.2

	12·60 5·20 0·76	Excavn. for cuttgs. E220 (paths at & (bellmouth
	12·60 5·20 0·15	Disposal of excvtd. E532 mat. (topsoil on (E.side

Banks

Width of banks (inc. 150 mm
additnl. excavn. for topsoil)

	E. side	W. side
CS1	150	2·650
CS2	850	2·850
	2)1·000	2)5·500
av. width	500 (fill)	2·750

		bellmouth
	850	2·850
	1·750	750
	2)2·600	2)3·600
	1·300 (fill)	1·800

height of banks

	E. side	W. side
CS1	150	1·050
CS2	450	950
	2)600	2)2·000
	300	1·000

		bellmouth
	450	950
	1·000	750
	2)1·450	2)1·700
	725	850

½/	30·30 2·75 1·00	Excavn. for cuttgs; (w.side Commg. Surf. u/s of topsoil. E220
½/	10·00 1·80 0·85	& (bellmouth (w.side Disposal of excvtd. mat. E532

14.3

The additional 40 mm depth of excavation over the areas of the two crossings is not large enough to justify separate measurement. Similarly the extra excavation for quadrants over that required for kerbs would be largely offset by the smaller quantity of excavation required for the granite setts – a sense of proportion must be maintained.

The build up of dimensions for the bank excavation is inserted in waste, to obtain the average widths and heights.
The topsoil component will require subsequent adjustment.
Disposal of excavated material is deemed to be disposal off the site unless otherwise stated in the item description (rule D4).

Slopes of 1 in 2 to banks have been assumed.

The volume of bank excavation = length x average width x average depth.

½/	30·30	**Fillg. embankments;**	Filling to embankments is
	0·50	selected excvtd. mat.	kept separate from general fill.
	0·30	other than topsoil or (E. side	The description must contain
½/	10·00	rock. E 624	the appropriate third division
	1·30		classification.
	0·73	(bellmth E. side	

	41·00	<u>Ddt.</u> Excavn. for cuttgs;	Adjustment of topsoil
	5·00	Commg. Surf. u/s of topsoil.	excavation over area of
	0·15	(carrgwy.	carriageway, paths and verges.
2/3/4/	10·70	& E220	The depositing and spreading of the topsoil will be picked up
	10·70		in subsequent verge and bank
	0·15	<u>Ddt.</u> Disposal of (bellmth.	slope items. The small surplus
2/	30·30	excvtd. mat. E532	quantity of topsoil can remain
	3·60	& (paths &	on the site and make up
	0·15	verges	surface irregularities.
2/	12·60	<u>Add</u> Excavn. for cuttings;	
	5·20	topsoil. E210	
	0·15	(bellmth.	

	Verges	The total lengths of verges
	30·300	are adjusted for the lengths of
	less crossings	the crossings.
	2·500	
	1·500 4·000	
	26·300	

2/	26·30	Fillg. thickness: 150 mm,	The soiling of verges is kept
	1·98	excvtd. topsoil. E641·1	separate from grass seeding. The grass seeding requirements
		&	will be extracted by the
		Landscapg., grass seedg.	Contractor from the Specification.
		E 830·1	

14.4

Banks

Three separate items arise in connection with the banks:
(1) trimming of slopes
(2) soiling of slopes
(3) grass seeding

27·70	Excvtn. ancills., trimming of excvtd. surfs., (W. side inclined at an L of 10° to 45° to the hor. (bell mth. w. side E512·1
2·85	
10·00	
1·90	

&

Fillg. thickness: 150mm, excvtd. topsoil; to surfs. inclined at an L of 10° to 45° to the hor. E641·2

&

Landscapg., grass seedg. to surfs. inclined at an L ex. 10° to the hor. E830·2

27·70	Fillg. ancills. trimmg of filled surfs., inclined at an L of 10° to 45° to the hor. (E. side E712·1
0·65	
10·00	
1·00	

(bell mth. E. side

&

Fillg. thickness: 150 mm excvtd. topsoil; to surfs. inclined at an L of 10° to 45° to the hor.

& E641·2

Landscapg., grass seedg. to surfs. inclined at an L ex 10° to the hor. E830·2

The filling item is measured in m^2 as it is to a stated depth or thickness; stating the appropriate inclination category from rule A14.

The grass seeding on banks has a separate classification to that of the verges, as it falls into the inclined category under rule A18.

The trimming of filled surfaces is kept separate from that to excavated surfaces, as the costs can vary significantly.

The same inclination categories apply as for filling to slopes.

Note the use of suffixes where standard descriptions are varied by the addition of further information.

Many details of road construction can be obtained from the Department of Transport 'Specification for Highway Works', to which reference can be made in the item descriptions. The actual thicknesses of slabs and courses can be given instead of the third division depth ranges in accordance with rule A1 of class R.

Roads and Pavings

41·00	Base granular mat. DTp Specfd. type I, depth: 75mm.
5·00	
	& R113
2/³⁄₁₄/	
10·70	Carriageway slab of (bell mth. DTp specfd. pavg. qual. jtd. reinfd. (JBC) conc., depth; 150mm. R414
10·70	

14.5

2/3/14/	41·00 5·00 10·70 10·70	Steel fabric reinft. to BS 4483, nom. mass 3-4 kg/m²; type A252. R443	Floating of channels and dishing around gullies are included in the carriageway slab rate (rule C1).

&

Waterproof membrane below conc. pavement; waterproof paper to BS 1521 class B1F.   R480

&

Excavn. ancillaries, prepn. of excvtd. surfaces.   E522·1

The waterproof membrane is likely to be of waterproof paper or impermeable plastic sheeting (250 or 500 grade).
Preparation of excavated or filled surfaces to receive permanent works is measured under class E (rules M11 and M23).

Joints in conc. pavements

5/ 2/	5·00 10·80 4·40 5·00	Expansion jts. depth: 150 mm; as detail J, Drawing 18/2 at 7·5 m centres. (transverse jts. (bellmth. (do. (do. R524

Expansion joints are always measured but construction joints only when they are expressly required (rule M7). No formwork is measured (rule C1).
The length of kerb is adjusted in 'waste' for the crossings on both sides of the road.
The kerb section is identified by reference to BS 7263. Kerbs laid to a radius exceeding 12 m are included with those laid straight. Details of concrete beds and backings to kerbs are included in the kerb descriptions (rule C3).

Kerbs, channels & edgings

less	30·200
setts 2·500	
quads. 2/300 = 600 3·100	
	27·100

2/	27·10	Precast conc. kerb to BS 7263 Pt.1, fig.1 (d), st. or curved to rad. ex. 12 m; bedded and backed w. conc. grade C10 as detail X; Dwg 18. R611

2/½/2/22/7/	10·70	Precast conc. kerb to BS 7263 Pt.1, fig.1 (d), curved to rad. n.e. 12 m; bedded & backed a.b. R612

The kerbs to the bellmouth are kept separate as they are laid to less than 12 m radius.

14.6

205

Granite setts

2/	2·50	
2/	10·70	
	5·00	

Granite sett edgings, (crossgs. 100 x 100 mm, st. or curved to rad. ex junctn. w xtg. road 12 m; bedded on conc. grade C10 as (do. detail Y, Dwg. 18.   R691

This item is not listed in class R and hence the figure 9 is used in the second division to represent a non-standard item.

2/	10·70	
	5·00	

Take up and remove xtg. precast conc. kerbs.
                                    R900

Kerbs at junction of new and existing roads; another non-standard item.

2/	2	

Precast conc. quadrant, 305 x 305 x 255mm type QHB to BS7263, fig. 1 (q) bedded & backed w. conc. grade C10 as detail Q, Dwg. 18/2.   R693

To crossings (one each side at junction of kerbs and setts). Enumerated item but following the same approach as for kerbs. Excavation was dealt with previously.

Light Duty Pavements
Crossings

		3·600
less setts		100
		3·500

Vehicular crossings traversing paths and verges.

2/	3·50	
	2·50	

Gran. base DTp Specfd. type 1, depth: 75 mm.
                                    R713
                    &

Similar base to that for carriageway.

In situ conc. to BS 5328 mix grade C15 depth: 100 mm; w. tamped non-skid fin.
                                    R773·1
                    &

Waterproof membrane below conc. pavement; w.p. paper to BS 1521 class B1F. R480

The description of the concrete slab follows the approach prescribed for light duty pavements, but substituting the grade of concrete in accordance with BS 5328.

14.7

2/	3·50		Excavn. ancillaries, prepn.
	2·50		of excvtd. surfs. E522.2

Footways
```
              30·200
less crossgs. 2·500
              27·700
```

```
        verge crossgs.
verge   2·100
less kerb 125
        1·975
```

2/	27·70		Gran. base, DTp specfd.
	1·50		type I., depth: 75 mm. R713

2/	1·98		& (verge crossgs.
	1·50		

Dense bit. macadam base-
course DTp specfd. clause 906

2/	12·60		depth : 50mm. (bellmth.
	5·20		R752

&

Dense bit. macadam wearg.
course DTp specfd. clause 912
depth : 10 mm.   R751

Edgings

2/	27·70		Precast conc. edging to
			BS 7263 Pt.1, fig. 1 (m), 50 x
2/	10·00		150 mm; st. or curved to rad.
			ex. 12 m, bedded & backed w.
			conc. grade C10 as detail E,
			Dwg. 18/2. R651

Surface Water Drainage
Depths (inc. 150 mm conc. bed)
```
gully      900
MH       1·287
       2)2·187
av. depth 1·094
```

14.8

Precast concrete flag descriptions include the types of slab in BS 7263 and the thickness.
Precast concrete edging is measured and described in a similar manner to precast concrete kerbs. Figure 1 (m) of BS 7263 shows three sets of dimensions for the round top variety and so dimensions have to be included in the description.
First calculate the average depth of the surface water gully connections.

The measurement of the preparation of formation and provision of waterproof membrane to crossings are kept separate from similar carriageway items, as they are likely to generate higher rates.
Each course constitutes a separate item and the particulars are obtained from the Department of Transport 'Specification for Highway Works', with the thickness given in each case.
Locational notes are given in waste for identification purposes.
All preliminary calculations are also inserted to prevent errors and provide the facility for checking.

	3·00		Clay pipes of S W qual. to BS 65 w. s & s flex. jts., nom. bore: 150 mm in trs., depth: ne 1·5 m; in rd. gully connectns. I 112·1	The descriptions of pipes include materials, joint types and nominal bores with references to British Standards where appropriate (rule A2 of class I).
	6·50			
			&	
			Surround, mass conc. grade C 10, pipe nom. bore: 150 mm; thickness: 150 mm. L 541·1	Materials and thicknesses of beds, haunches and surrounds are stated in item descriptions (rule A3 of class L).
2/	1		Clay pipe fittgs. S W qual. to BS 65 w. s & s flex. jts., bends, nom. bore: 150 mm. J 111	Pipe fittings are enumerated giving similar particulars as for pipes (rule A1 of class J).
2/	1		Gullies, precast conc. trapped; to BS 5911, fig. 2(a) as detail 2 Dwg. 18, w. Broads Nr. 205 kerb inlet gully cover. K 360·1	Gullies are enumerated with adequate references for detailed particulars and stating the type of cover (rules A1 and A2 of class K).

14.9

13 Measurement of Pipework

The rules for the measurement of pipes and associated work occupy four classes of the CESMM3 Work Classification (I-L). These are all closely interrelated and should be considered as a composite class.

The rules in Class I cover pipework, Class J deals with pipe fittings and valves, Class K embraces manholes and work associated with pipework such as land drains, ditches, culverts, crossings and their reinstatement, while Class L is concerned with work related to the laying of pipes, such as extra cost items in trenching, bedding, haunching, surrounding, wrapping and pipe supports.

Pipework (Class I)

Pipework is measured under Class I in metres giving the nominal bore and trench depth ranges, although it is much more realistic to give the actual bore of the pipes as required by rule A2. Pipework items are comprehensive ones in that they include the following items in addition to the provision, laying and jointing of pipes.

(1) Jointing material (rule A2)
(2) Cutting of pipes (rule C1)
(3) Lengths occupied by fittings and valves and those built into chamber walls (rules M3 and M5)
(4) Excavation of trenches (rule C2)
(5) Backfilling of trenches with excavated material (rule C2)
(6) Upholding sides of excavation (rule C2)
(7) Preparation of surfaces (rule C2)
(8) Disposal of excavated material (rule C2)
(9) Removal of dead services (rule C2).

Although the classification table of Class I covers 512 different categories through classifications I 111 to I 888, nevertheless they cannot embrace every conceivable alternative. Hence rule A2 expands the first division rules to include separate items for different nominal bores, pipe materials, joints and linings. Work is also to be separated locationally to take account of different working conditions with resulting variations in cost (rule A1), paying particular attention to work in roads, through back gardens, restricted access and working conditions, need or otherwise for trench support and working around existing services.

CESMM3 incorporates total trench depths in pipework descriptions. Probably the most satisfactory approach is to enter pipe and locational descriptions in a heading, followed by the linear items subdivided between the various trench depth ranges. The prescribed depth zones can be readily determined on a longitudinal section by preparing a scale or strip containing the total depths drawn to the vertical scale used for the drawing. Sliding this scale along the drawing will identify the depth stages, which can be marked and the pipe runs broken down into lengths appropriate to each depth range. On steep gradients the length of pipe to be measured will be greater than the length on plan.

Where more than one pipe is to be laid in the same trench the item descriptions should indicate the use of a shared trench (rule A5).

Where a short pipe run not in a trench contains numerous valves and fittings, the length of pipe occupied can be excluded (rule M3). Where pipe bores change at fittings or valves, the different pipe lengths should be measured to the centre of the fitting or valve. Separate items are to be provided for pipes that are not laid in trenches, as amplified by rule D1.

Fittings and Valves (Class J)

All fittings and valves are enumerated with a full description in each item. Separate items are not required for excavation, preparation of surfaces, disposal of excavated material, upholding sides of excavation, backfilling and removal of dead services (rule C2).

The item descriptions for pipe fittings, such as bends, junctions and tapers are to include nominal bore, material, jointing and lining particulars, and reference made to applicable British Standards (rule A1). In general, lengths and angles of bends, junctions and branches need not be stated, except for cast iron or spun iron fittings exceeding 300 mm nominal bore and all steel fittings (rule A2). The latter fittings are expensive and the inclusion of principal dimensions such as the effective length, nominal bore and angle of a bend, will permit identification in the supplier's catalogue. Rule A6 also requires additional particulars in item descriptions of valves and penstocks.

The list of fittings and valves in the second division is not intended to be exhaustive and only gives the most commonly encountered components. Others can be added and coded as J*9*. Where pipe fittings, such as branches and tapers, cross the third division nominal bore ranges, they shall be classified in the larger size range (rule D1).

Straight specials are pipes either cut to length or made to order (non-standard lengths) (rule D2) and are measured only when express-ly required (rule M2).

Manholes and Pipework Ancillaries (Class K)

Manholes, other chambers and gullies are not measured in detail but are enumerated in accordance with the principles listed in K1–3 and rules C3, C4, A1 and A2. They are identified by a type or mark number, which will be the reference for constructional details given in Drawings and Specification, and are deemed to include all items of metalwork and pipework, other than valves (rule C3). Differing manhole and other chamber arrangements will be evidenced by separate bill items containing different type or mark numbers. Excavation in rock or artificial hard material and backfill with other than excavated material is measured as an extra item in Class L. Manholes with backdrops are separately classified so that the backdrop pipe will be included in the manhole item and not the pipework, although the latter is normally taken to the inside face of the manhole (rule C4). Complex or unusual manholes may be measured in detail as example VII (footnote to page 53 of CESMM3).

French and groundwater drains incorporating pipes will be meas-ured under Class I, including excavation, backfill with and disposal of surplus excavated material. Backfill with porous material is, however, measured separately in cubic metres (K 4 1–2), stating the nature of the filling material (rule A4). Where no pipes are to be laid, the trench preparation is measured in metres (K 4 3).

Ditches are measured in metres stating the cross-sectional area in the ranges listed in the third division. A lined ditch description shall include the nature and dimensions of the lining (rule A5).

Pipe crossings of streams are not measured where the width does not exceed 1 m (rule M5). Crossings of hedges, walls, fences, sewers or drains, and other stated underground services are separately enumer-ated, giving the appropriate pipe bore range in the description.

The measurement of crossings and reinstatement does not dis-tinguish between different pipe materials or depths, nor is special mention made of pipes in shared trenches or of manholes. Rule A8 requires the types and depths of surfacing to be stated in reinstate-ment items. The reinstatement classifications distinguish between grassland, gardens, sports fields and cultivated land (rule A10).

Timber and metal supports left in excavations (measured in m² of the supported surface) are only taken where it is a requirement of the Engineer (rule M11). Item descriptions of connections of new pipework to existing work shall identify the nature of the existing service and will include details of associated work such as sustaining flows and reconstructing benching in manholes (rule A12).

Supports and Protection, Ancillaries to Laying and Excavation (Class L)

Class L covers the measurement of activities related to excavating and backfilling pipe trenches and associated work which give rise to extra cost, pipelaying in headings and by thrust boring and pipe jacking, and for supporting and protecting pipework.

Where excavation and backfill of pipe trenches involve other than ordinary soft material, such as rock, mass concrete, reinforced concrete or other artificial hard material, then an item in m³ is required to cover the 'extra cost' of dealing with these materials over and above the pipework items already measured in Class I. The quantity is computed by multiplying the average length, average depth and nominal width of trench excavation stated in the Contract, making allowance for battered trench sides (rule M4). Where no nominal width is stated it shall be taken as 500 mm greater than the maximum nominal distance between the internal faces of the outer pipe walls where this distance does not exceed 1 m, and as 750 mm greater than this distance where it exceeds 1 m (rule D1). An isolated volume of hard material shall not be measured separately unless its volume exceeds 0.25 m³ (rule M8). Excavation which is expressly required to be carried out by hand shall form separate items (rule A1).

Backfilling above the Final Surface shall only be measured where the Engineer will not permit the use of excavated material, while excavation and backfilling below the Final Surface shall be measured only when required by the Engineer, for example 'soft spots' but not excess excavation (rule M7).

Similarly pipe laying in headings and by thrust boring and pipe jacking shall be measured only when expressly required (rule M9) and shall give the location so that it can be correlated with pipework items in Class I (rule A2). The plant and temporary works associated with thrust boring and pipe jacking will probably generate method-related charges. Pits for thrust boring and pipe jacking may either be 'specified requirements' in Class A or method-related charges where at the discretion of the Contractor.

Lengths of support and protection to pipelines shall include fittings and valves but exclude manholes and chambers that provide breaks in the support or protection (rule M11). The materials used shall be stated but not the cross-sectional dimensions, which are obtainable from the Drawings. Nominal bore ranges are given in the third division, although the actual nominal bore can be given where only one pipe size is involved (5.14 of CESMM3). The bed items include excavation and they can be combined with haunches or surrounds where they are of the same material (rules C1, D2 and A3). Wrapping and lagging of pipes are measured in metres, including the lengths occupied by fittings and valves (rule M12).

Concrete stools and thrust blocks are enumerated and descriptions include the type of concrete and appropriate range of concrete volume, with no separate measurement of formwork or reinforcement (rule C2). Other isolated pipe supports are also enumerated but stating the principal dimensions, materials and the height measured in accordance with rule D5.

Worked Examples

Worked examples follow covering the measurement of a sewer and a water main.

Example XV (Drawings Nos. 19 and 20) — Sewer

It is desirable and quicker, and there is less risk of error if sewer and manhole schedules are prepared on the lines indicated in this example. The 'taking-off' process then becomes greatly simplified, merely involving the extraction of the particulars from the schedule and combining totals where appropriate.

For layout reasons, and for ease of reading, this page has intentionally been left-blank.

Sewer Schedule

Location	Type and Size of Pipe	Length of Pipe in m		Nr & Size of Junctns.
MHs 1–2	225 mm conc. pipe	less mhs	90.000 1.200	—
			88.800	
2–3	225 mm conc. pipe		162.000 90.000	—
			72.000	
		less mhs	1.200	
			70.800	
3–4	150 mm gvc pipe		205.000 162.000	—
			43.000	
		less mhs	1.200	
			41.800	
4–5	150 mm gvc pipe		265.000 205.000	4 nr 100/150 mm
			60.000	
		less mhs	1.200	
			58.800	
5–6	150 mm gvc pipe		321.000 265.000	4 nr 100/150 mm
			56.000	
		less mhs	1.200	
			54.800	
6–7	150 mm gvc pipe		387.000 321.000	—
			66.000	
		less mhs	1.200	
			64.800	

Length of Trench 1.5–2 m (Total Depth)	Length of Trench 2–2.5 m (Total Depth)	Length of Trench 2.5–3 m (Total Depth)	Length of Trench 3–3.5 m (Total Depth)	Length of Trench 3.5–4 m (Total Depth)	Nr of Hedge and Fence Crossings
—	—	16.300 (field)	14.000 (field)	6.500 (road) 10.000 (verge) 42.000 (field)	1 fence 1 hedge
—	70.800 (field)	—	—	—	2 hedges
35.800 (field) 6.000 (road)	—	—	—	—	1 hedge
58.800 (road)	—	—	—	—	—
54.800 (road)	—	—	—	—	—
13.000 (road) 51.800 (verge)	—	—	—	—	—

Manhole Schedule

Manhole nr	Ground Level	Invert Level	Total Depth + 170 mm (for base and channel)	Depth of 1200 mm Rings	Depth of 1200– 675 mm Taper	Depth of 675 mm Rings
1	240.000	236.000	4.000	1.500	600	1.000
2	238.800	236.300	2.500	750	600	300
3	238.500	236.540	1.960	600	600	—
4	238.650	236.755	1.895	450	600	—
5	239.000	237.055	1.945	600	600	—
6	239.000	237.335	1.665	300	600	—
7	239.400	237.665	1.735	300	600	—
Totals	—	—	15.700	4.500	7 nr (4.200)	1.300

Depth of 215 mm bwk.	Type of m.h. Cover	Sewer Sizes	Junctions Size & nr	Type of Channel	Nr of Step Irons (all in pre-cast units)	Combined depth of Cover, Cover Slab & Base Wall	Location
150	Medium	2/225	—	225 straight	10	750	field
100	Medium	2/225	—	225 curved	5	750	field
—	Medium	1/225 1/150	—	225–150 straight	4	750	field
95	Heavy	3/150	1/150	150 curved with junctn.	4	750	road
—	Heavy	2/150	—	150 curved	4	750	road
15	Heavy	2/150	—	150 curved	3	750	road
85	Heavy	2/150	—	150 straight	3	750	road
445	3 Med. 4 Heavy	5/225 10/150	1/150	—	33 nr	—	

SEWER

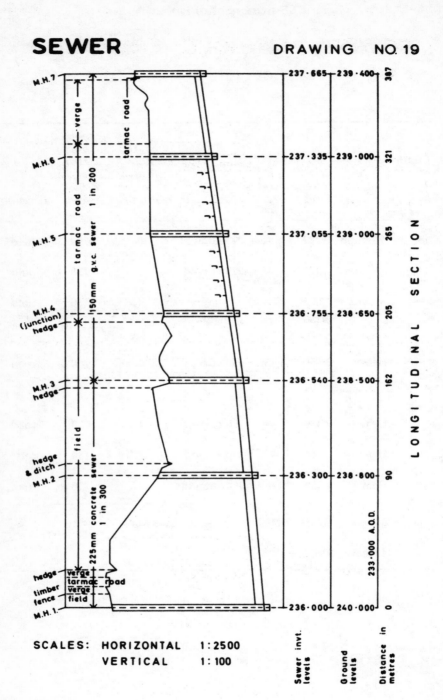

LONGITUDINAL SECTION

SCALES: HORIZONTAL 1:2500

VERTICAL 1:100

220

SEWER MANHOLE DETAILS

DRAWING NO. 20

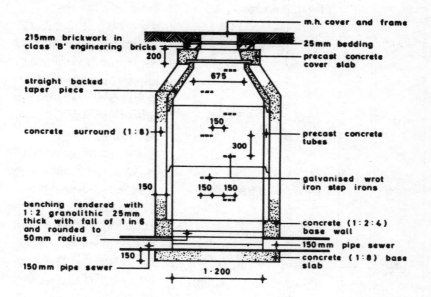

215mm brickwork in class 'B' engineering bricks

200

straight backed taper piece

concrete surround (1:8)

150

benching rendered with 1:2 granolithic 25mm thick with fall of 1 in 6 and rounded to 50mm radius

150mm pipe sewer

m.h. cover and frame

25mm bedding

precast concrete cover slab

675

150

300

precast concrete tubes

galvanised wrot iron step irons

150   150

concrete (1:2:4) base wall

150 mm pipe sewer

concrete (1:8) base slab

150

1·200

SECTION

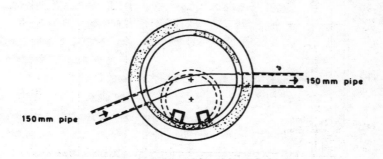

150 mm pipe

150 mm pipe

PLAN

SCALE   1:50

221

| SEWER | | | EXAMPLE XV |

SEWER

EXAMPLE XV

PIPEWORK — PIPES

<u>Conc. pipes of class M to BS 5911 w. s & s flexible jts., nom. bore: 225 mm in trenches.</u>
<u>Between manholes 1 & 3</u>

70·80	Depth: 2 – 2·5 m in field. (mhs 2–3) I 224.1
16·30	Depth: 2·5 – 3 m in field. (mhs 1–2) I 225.1
14·00	Depth: 3 – 3·5 m in field. (mhs 1–2) I 226.1
6·50	Depth: 3·5 – 4 m in road. (mhs 1–2) I 227.1
10·00	Depth: 3·5 – 4 m in verge. (mhs 1–2) I 227.2
42·00	Depth: 3·5 – 4 m in field. (mhs 1–2) I 227.3

<u>Clay pipes of normal quality to BS 65 w. s & s flexible jts., nom. bore: 150 mm in trenches.</u>

<u>Between manholes 3 & 7</u>

35·80	Depth: 1·5 – 2 m in field. (mhs 3–4) I 113.1

The description of the pipes shall include materials, joint types, nominal bores and lining requirements, with references to British Standards where appropriate (rule A2 of class I). The locations of pipe runs shall also be given for identification purposes (rule A1) and differing working conditions resulting in different costs shall also be identified.

Pipes in trenches are subdivided into the total depth ranges entered in the third division of class I. The pipe items include excavation, preparation of surfaces, disposal of excavated material, upholding sides of excavation, backfilling and removal of dead services (rule C2).

Lengths are measured up to the inside surfaces of manhole walls (rule M5).

Any concrete or other protection to pipes would be measured under Class L.

The change of pipe material and bore entails a new heading, followed by the appropriate locational reference.

All the necessary particulars can be extracted direct from the sewer schedule.

15.1

SEWER	(Contd.)
6·00	Depth: 1·5 – 2m (mhs 3-4 in road.
58·80	(mhs 4-5
54·80	(mhs 5-6
13·00	(mhs 6-7
	I 113·2
51·80	Depth: 1·5 – 2m (mhs 6-7 in verge. I 113·3

PIPEWORK – FITTINGS & VALVES

Clay pipe fittings to BS 65 w. s & s flex. jts.

| 4 | Junctions & branches: nom. bore: 150 mm. (mhs 4-5   J 121 |
| 4 | &   (mhs 5-6 Stoppers; nom. bore: 150 mm.   J 191 |

PIPEWORK – MANHOLES AND PIPEWORK ANCILLARIES

Manholes

| 1 | Precast conc. depth : 1·5 – 2m; type CMH1 (Dwg. 20) w. c.i. cover to BS.497 ref. (mh 3 MB2 - 50   K 152·1 |
| 4 | Precast conc. depth : 1·5 – 2m; type CMH1 (Dwg. 20) w. c.i. cover to BS.497 (mhs 4,5,6,7 ref. MA - 50   K 152·2 |

Continue with the measurement of the pipes and fittings until all have been entered. Double check against the schedule entries and cross out each item as transferred to the dimensions sheets. Similar provisions apply to fitting descriptions as for pipes (rules C1, C2 and A1 of class J).

The actual bores of clay pipe fittings should be stated, as prescribed by rule A1 of class J. Where fittings comprise pipes of different nominal bores, they shall be classified in the third division by the nominal bore of the largest pipe (rule D1 of class J).

Manhole types or mark numbers are to be included in item descriptions and the constructional particulars are obtained from the detailed drawings. Separate items are required for manholes of different materials, in different depth ranges and with different covers, but manhole items are deemed to include all items of metalwork and pipework, other than valves (rule C3 of class K).

15.2

			SEWER		(Contd.)	

1	Precast conc. depth: 2 - 2·5 m; Type CMH1 (Dwg. 20) w. c.i. cover to BS 497 ref. (mh 2 MB2 - 50 K153	Alternatively, each manhole could be allocated a separate type number on the manhole schedule, but this would result in numerous separate items of similar manholes which is clearly not the intention of CESMM3. The manhole schedule will assist the tenderer in pricing.
1	Precast conc. depth: 3·5 - 4m; Type CMH1 (Dwg. 20) w. c.i. cover to BS 497 ref. (mh 1 MB2 - 50 K156	
	Crossings	The ditch crossing is not taken as it is assumed not to exceed 1 m wide (rule M5 of class K).
4	Hedges ; pipe nom. bore: n.e. 300 mm. (mhs 1-4 K641	It is <u>not</u> necessary to classify separately pipes of differing nominal bores if within the same pipe bore range in the third
1	Fence ; pipe nom. bore: n.e. 300 mm. (mhs 1-2 K661	division, as they will have little affect on price.
	Reinstatement	A combined item of breaking up and reinstatement is required for paved surfaces measured in
6·00	Breakg. up & (mhs 3-4 tempy. reinstatement	metres on the lines of the pipe trenches. The type and maximum
58·80	of rds.; pipe nom. (mhs 4-5 bore: n.e. 300 mm;	depth of surface is stated in the description, but no item is
54·80	flexible rd. constn. (mhs 5-6 max. depth: 75 mm.	required for removal and rein-statement of kerbs (rule C8 of
13·00	w. 200 mm (mhs 6-7 sub - base.	class K).
6·50	(mhs 1-2	An additional length is given to pick up reinstatement over the
4/ 1·20	(mhs 4,5 6&7 K711	manholes, which is excluded from the pipe lengths. Pipe ranges are adequate for this class of work.

15.3

		SEWER	(Contd.)		

	35·80	Reinstatement of grassland;	(mhs 3-4	Reinstatement of roadside verges has been kept separate from
	16·30	pipe nom. bore: n.e. 300 mm.	(mhs 1-2	grassland to allow the tenderer to price it differently, and as this
	14·00		(mhs 1-2	is not listed in the second division, the digit 9 is inserted
	42·00		(mhs 1-2	in the code.
	70·80		(mhs 2-3	Testing of pipes would be covered by a General Item (A260).
3/	1·20		(mhs 1,2 & 3 K751·1	

	51·80	Reinstatement of roadside verge;	(mhs 6-7	Check the sewer schedule to ensure that all lengths have
	10·00	pipe nom. bore: n.e. 300 mm.	(mhs 1-2 K791	been taken. The omission of quantities is a serious matter.

	4	Marker posts; 50 x 50 x 600 mm	(mhs 4-5	The marker posts are needed to indicate the positions of the
	4	oak set in conc. base.	(mhs 5-6 K820	junctions. Item descriptions are to include sizes and types of posts (rule A11 of class K).

15.4

WATER MAIN

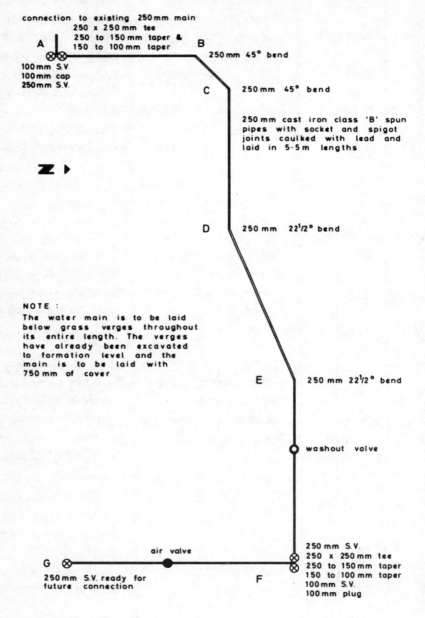

connection to existing 250 mm main
250 x 250 mm tee
250 to 150 mm taper &
150 to 100 mm taper

A **B**

250 mm 45° bend

100 mm S.V.
100 mm cap
250 mm S.V.

C 250 mm 45° bend

250 mm cast iron class 'B' spun
pipes with socket and spigot
joints caulked with lead and
laid in 5·5 m lengths

D 250 mm 22½° bend

NOTE :
The water main is to be laid
below grass verges throughout
its entire length. The verges
have already been excavated
to formation level and the
main is to be laid with
750 mm of cover

E 250 mm 22½° bend

washout valve

air valve

250 mm S.V.
250 x 250 mm tee
250 to 150 mm taper
150 to 100 mm taper
100 mm S.V.
100 mm plug

G **F**

250 mm S.V. ready for
future connection

SCALE   1:1250

EXAMPLE XVI

PIPEWORK — PIPES

<u>Cast (spun) iron s&s
pipes to BS 1211 (class B)
with caulked lead jts.; nom.
bore: 250 mm</u>
(A - G on Dwg. Nr. 21)

7·50	In trenches,	(to A
	depth: n.e. 1·5 m.	
47·50		(A - B
16·00		(B - C
48·50		(C - D
56·50		(D - E
63·50		(E - F
75·00		(F - G
		I322

PIPEWORK FITTINGS & VALVES

Cast (spun) iron s&s
pipe fittgs. to BS 1211
(class B) with caulked
lead jts.

2/	I	Taper; nom. bore: (A & F
		150 – 250 mm.
		J332
2/	I	Taper; nom. bore: (A & F
		100 – 150 mm.
		J331

The material, joint type and
nominal bore shall be given with
references to British Standards
where appropriate (rule A2 of
class I).
The location of each item or
group of similar items is to be
given, so that the pipe runs can
be identified by reference to the
drawings (rule A1).
Lengths of pipes are measured
on their centre lines, including
the lengths occupied by fittings
and valves (rule M3).
This item includes the provision,
laying and jointing of pipes in
trenches and excavating,
preparing surfaces, disposal of
excavated material, upholding
sides of excavation, backfilling
and removal of dead services
(rule C2).
Trench depth ranges are
measured from Commencing
Surface to inverts of pipes
(rule D3).

Fittings are enumerated giving
similar particulars as for pipes.
Arranging all the items under
appropriate headings reduces the
length of entries and groups
items ready for billing.
Location references are given in
waste for identification purposes.

16.1

WATER MAIN (Contd.)

2/	1	Junctions & branches; nom. bore: 250 x 250 x 250 mm.	(A & F J322.1
	1	Junctions & branches; nom. bore: 250 x 250 x 63 mm w. flanged branch.	(WO E-F J322.2
2/	1	Bends; nom. bore: 250 mm, 45°.	(B & C J312.1
2/	1	Bends; nom. bore: 250 mm, 22½°.	(D & E J312.2
3/	1	Flanged spigots; nom. bore: 250 mm. J352.1	
		&	
		Flanged socs; nom. bore: 250 mm.	(A.F & G J352.2
2/	1	Flanged spigots; nom. bore: 100 mm. J351.1	
		&	
		Flanged socs; nom. bore: 100 mm. J351.2	(A & F
	1	Plug nom. bore: 100 mm.	(F J391.1
		&	
		Cap nom. bore: 100 mm.	(A J391.2

16.2

The fittings items are really 'extra over' items as the pipes, with their accompanying excavation and backfill have already been measured for the lengths occupied by fittings. Vertical bends in metal pipework exceeding 300 mm bore have to be so described (rule A3 of class J). In strict compliance with CESMM3, the angles of bends and effective lengths would be given where the nominal bore exceeds 300mm. Hence the addition of angles in these items is not a requirement of the CESMM, but is included as being of assistance to the estimator. Note use of suffixes to cover additional descriptions.

Flanged spigots and sockets are needed to connect spigot and socket pipes to the flanged valves. It is generally considered advisable to use sluice valves with flanged joints as they can withstand more effectively the pressures resulting from the opening and closing of the valves. These items have been coded under adaptors (category 5 in second division). Alternatively they might be given the independent classification of 9, i.e. J391.

These fittings are not listed in the second division of class J and hence the digit 9 is used in the coding (second division).

			Description	Notes
			<u>Cast iron gate valves hand operated to BS5150, class 125, clamp patt., w; extensn. spindles, flanged jts., & T keys.</u>	The descriptions of gate valves shall include such particulars as materials, nominal bores, joints and extension spindles, and provide references to applicable British Standards and specified qualities (rule A6 of class J).
3/	1		250 mm nom. bore. (A.F&G J812·1	
2/	1		100 mm nom. bore. (A&F J811·1	
			<u>Cast iron air valve, small single orifice as X catalogue Nr. 2416.</u>	The remaining valves are enumerated with essential particulars listed in accordance with rule A6.
	1		J861 (F-G	
			<u>Cast iron hydrant as washout, spindle type as X catalogue Nr. 347C.</u>	The supplier's catalogue reference can be given as an alternative to a British Standard.
	1		J891 (E-F	

PIPEWORK - VALVE CHAMBERS
AND PIPEWORK ANCILLARIES

			Description	Notes
			<u>Valve chambers brick depth: n.e. 1·50 m; type VCI</u>	It is considered more appropriate to describe these as valve chambers, as they come within the other stated chambers category. A drawing will show the detailed construction for a type VCI chamber and it does not therefore require repeating in the item description.
5/	1		w. c.i. SV cover as X catalogue Nr. 47A. K211·1	

16.3

WATER MAIN (Contd.)

Valve chambers (contd.)

1	w. c.i. W.O. cover as X catalogue Nr. 51B.

              &   K211·2

w. c.i. A.V. cover as
X catalogue Nr. 52C.

            K211·3

The suffixes are added to the code numbers relating to the valve chamber items, to signify the different covers.

The measurement of the water main finishes with the relevant items of other pipework ancillaries in K8.

Other pipework ancillaries

7/   1   Marker posts; r. conc. (p.c.£12·50 ea.), bolting on plate supplied by Water Authority & settg. in conc. as detail E, Dwg. WS 26.

           K820

Valve markers require additional particulars to enable them to be priced, involving where appropriate references to Drawings and/or Specification.

1   Connectn. of pipe to xtg. pipe, nom. bore 250 mm, inc. excavn. to locate pipe & burning out plug.

           K862·1

Enumerated connection item incorporating any additional particulars considered necessary.

The testing of the pipeline should be included in the General Items (class A) under classification A260, stating the length and nominal bore of pipe, test pressure and period of test.

16.4

14 Measurement of Tunnels

The cost of tunnelling is influenced greatly by the nature of the material to be excavated and supported, and because of its relative inaccessibility before work is under way, it cannot be assessed accurately at tender stage. The work is highly mechanised and therefore extremely expensive, but is also subject to severe constraints owing to limitations of access and area of working face, and the uncertainty of ground conditions.

The cost uncertainty is generally greater than with other types of civil engineering work and the rules of measurement take this into account by limiting the risk borne by the Contractor, principally through the measurement of compressed air working, temporary support and stabilisation.

Thus rule A1 prescribes that work expressly required to be executed under compressed air shall be measured separately stating the gauge pressure in stages. Items are also to be included as specified requirements under Class A for the provision and operation of plant and services for this work. Hence the responsibility for deciding the extent of compressed air work and the operative air pressure rests with the Engineer when formulating the Bill. Any deficiencies will be subsequently rectified through variation orders.

Rule M8 prescribes that 'both temporary and permanent support and stabilisation shall be measured', and these are not restricted to the normal 'expressly required' provision. Thus the Contractor will be reimbursed for the amount of support and stabilisation he provides at the billed rates, regardless of the quantity inserted in the Bill. This work includes installing rock bolts or steel arches, erecting timber supports, lagging between arches or timbers, applying sprayed concrete or mesh or link support, pressure grouting and forward probing. Hence most of the risk to the Contractor arising from the extent of support and stabilisation is removed.

Some engineers fear that contractors will be tempted to undertake more support work than is really justified, having inserted favourable rates in the Bill. There are however safeguards built into the contractual arrangements: the Engineer is empowered by the ICE Conditions of Contract to supervise the construction of the works and also the Contractor when pricing the Bill is in competition with other tenderers. Furthermore there is little incentive for the Contractor to slow down the driving operation to carry out unnecessary support work.

Barnes[5] recommends that tunnelling should be suitably divided into separate bill sections, in accordance with paragraph 5.8 of CESMM3, because of the significant effect of location and construction method on tunnelling costs.

The excavation of tunnels and shafts is kept separate as is also that in rock from other stated material. An excavated surface measurement in m^2 is taken for the pricing of excavation and disposal of overbreak and subsequent back grouting to fill voids. Excavation rates are deemed to include disposal of excavated material off the site unless otherwise stated in item descriptions (rule C1). The volume of excavation, and areas of excavated surfaces and *in situ* linings are based on payment lines shown on the drawings, while any cavity formed outside these payment lines is deemed to be overbreak. Where no payment lines are shown on the drawings, the overbreak starts either at the limit of the permanent work to be constructed in the tunnel or shaft or at the minimum specified size of the void required to accommodate the permanent work (rules M2, M4 and M5). Separate items for excavation and linings are needed for curved or tapered tunnels and shafts, tunnels sloping at 1 in 25 or steeper, and inclined shafts (rule A3). The diameter used for classification and inclusion in item descriptions of excavation for tunnels, shafts and other cavities shall be the external diameter (rule D2).

Preformed segmental linings are measured by the number of rings, since both the cost of lining materials and of labour is proportional to them, and this eliminates the need to make an accurate prior assessment of creep. Item descriptions for preformed segmental linings are to state the nominal ring width and list the components of one ring of segments, often including the number of bolts, grummets and washers, and the maximum weight of a piece of segmental lining (rule A9).

Barnes[5] describes how pressure grouting (T83*) refers to the treatment used to support and stabilise the ground surrounding a tunnel and thus constitutes a distinct and separate item from back grouting of voids resulting from overbreak.

Worked Example

A worked example follows covering the measurement of a cast iron tunnel lining.

For layout reasons, and for ease of reading, this page has intentionally been left-blank.

CAST IRON TUNNEL LINING

DRAWING NO. 22

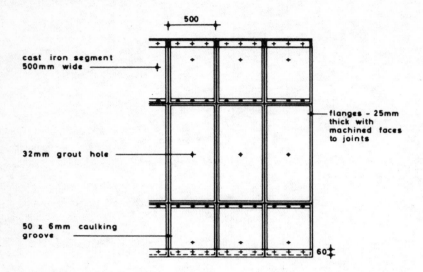

500

cast iron segment
500mm wide

flanges - 25mm
thick with
machined faces
to joints

32mm grout hole

50 x 6mm caulking
groove

60

INTERNAL ELEVATION

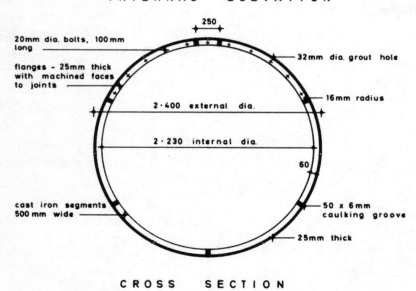

250

20mm dia. bolts, 100mm
long

32mm dia. grout hole

flanges - 25mm thick
with machined faces
to joints

16mm radius

2·400 external dia.

2·230 internal dia.

60

cast iron segments
500 mm wide

50 x 6mm
caulking groove

25mm thick

CROSS SECTION

SCALE 1:40

BRADWELL SEWER TUNNEL

350 m length
Tunnel excavation
ext. diameter 2·40 m

22/7	350·00 1·20 1·20	In med. dense gravel & sand ; st. T122	A locational heading is needed for purposes of identification. The diameter stated is the external diameter of the tunnel lining. The material through which the tunnel is to be driven is stated. If the work was under compressed air this would be stated in a heading, with the operative gauge pressure range (rule A1 of class T). As no payment lines are shown on the Drawing, the excavated surface is measured to the area enclosing the permanent work (net void). The Contractor will price the over-break (excess excavation) and subsequent grouting in this item (rule M4).

22/7	350·00 2·40	Excavated surfs. in med. dense gravel & sand; voids filled w. ct. grout as Spec. Clause 184. T180	

Segmental lings.

$$500\ )\ \overline{350\cdot000}$$
$$\overline{700}$$

bolts/ring

circumf. 6×5+1	31	
long. 7×3	21	
	52	

700/	1	Preformed segmental tunnel lings. of c.i. bolted rings, ext. diameter 2·40 m ; ring nom. width 500 mm, comprisg. 7 segments, max. piece wt. 144 kg, 52 bolts & grummets & 104 washers. T532	Number of bolts per ring are calculated in waste, to ensure accuracy and permit checking. Preformed segmental linings are enumerated by rings, giving the nominal ring width, number of segments, bolts, grummets and washers, and maximum piece weight (rule A9). The measurement unit for packing shall be the number of rings of segments packed (rule M7).

&

Parallel circumferential packg. for preformed segmental tunnel lings; bit. impregnated. fibrebd., thickness : 8 mm.
    T571

caulkg. jts.

ling. thickness	25
½ flange depth – ½/60	30
	55

Build up of caulking dimensions in waste, to arrive at the mean length to be caulked.

ext. diam. of ling.   2·400
less 2 x outer face of ling.
to centre of caulkg.
groove. 2/55   110
mean diam. of groove   2·290

17.1

TUNNEL LINING (Contd.)

700/22/7	2·29	Lead fibre caulkg. (circum.) for preformed segmental tunnel (long. lings. T574	Caulking of grooves between segments to ensure watertight joints is measured as a linear item.
700/7/	0·50		

If pressure grouting is required it will be measured in the following manner :

This may be required to support and stabilize the ground surrounding the tunnel.

Pressure grouting

	1	Sets of drillg. & groutg. plant. T831	An enumerated item for drilling and grouting plant.
700/	7	Face packers. T832	There is one grout hole to each segment. The lengths of holes are stated in stages of 5 m in item descriptions for drilling and re-drilling holes for pressure grouting (rule A17).
700/	4·00	Drillg. & flushg., hole diam. 32 mm, len. ne 5 m. T834	
700/	2·00	Re-drillg. & flushg. holes, len. ne 5m. T835	
	450 t	Injectn. of ct. grout as Spec. clause 215. T836	The quantity of cement grout has been calculated on the basis of 75 mm average thickness around the lining.

The following additional support and stabilization item might be needed :

	200·00	Fwd. probg., len. : 5 - 10 m. T840	A linear item with lengths of holes stated in 5 m stages (rule A17).

17. 2

236

15 Measurement of Rail Track

Since the materials for rail track may be supplied by the Employer, supplying and laying track are measured separately. Track foundations alone are covered by items that include both supplying and laying. 'Supply' includes delivery of components to the site (rule C3), while 'laying' comprises all work subsequent to delivery of components to the site (rule C8). Where track is not to be supplied by the Contractor, the location is to be stated in accordance with rule A13 of Class S.

Separate items are required for bottom ballast, placed before the track is laid, and top ballast, which is placed after the track is laid. The volume of top ballast includes the volume occupied by the sleepers (rule M1). The ballast rates must allow for the cost of boxing up, trimming to line and level and tamping after the track has been laid.

Enumerated items for the supply of sleepers shall state the type, size and identify fittings attached by the supplier (rule A8). Item descriptions for the supply of rails shall give the section reference or cross-sectional dimensions and the mass/metre of the weighted rail (rule A9). Item descriptions for turnouts and crossings shall state the type and shall be deemed to include timbers, fittings and check rails (rules A10 and C6), while the enumerated items for chairs, base plates, fishplates (in pairs) and related items are deemed to include fixings, keys, clips, bolts, nuts, screws, spikes, ferrules, track circuit insulators, pads and conductor rail insulator packings (rule C5).

It is unnecessary to give lengthy descriptions of materials for linear track laying items, the type and mass/metre of rail and type of joint and sleeper only being required by rule A15. Laying plain track is measured along the centre line of the track (two rail) and includes sleepers and fittings (rule C9), and shall include the lengths occupied by turnouts and diamond crossings (rule M8). Forming curves in plain track are separated according to the radius (not exceeding and exceeding 300 m), and constitute 'extra over' straight track items to pick up the additional costs. The term 'plain track' denotes track consisting of ordinary lengths of running rails. Item descriptions for laying turnouts and diamond crossings shall state their type and length (rule A16) and for buffer stops shall include their approximate weight (rule A18).

The classification of S 2 ** covers taking up existing track, with the track measured in metres and other items enumerated. Item descriptions shall state the amount of dismantling, details of disposal of track, and the type of rail, sleeper and joint (rule A3). Another classification (S 3 1–5 0) provides for the measurement of lifting, packing and slewing existing track, measured by number, but stating the length of track (rule D3), maximum distance of slew and the maximum lift (rule A5). This entails separate items for dealing with different lengths.

Worked Example

A worked example follows covering the measurement of railway track.

Example XVIII — Railway Track

150 m length of straight single track; excavation assumed to have been measured.

Specification Notes

(1) *Ballast*. The ballast shall be clean, hard broken stone to pass a 60 mm ring, laid to a width of 3.15 m. The bottom ballast shall be laid after the formation has been prepared and rolled to a depth of 450 mm below top of rail level.

Before the sleepers are laid, bottom ballast shall be laid to a consolidated depth of 150 mm. The permanent way material shall then be laid and the sleepers packed up with the top ballast for a width of 375 mm on each side of each rail. After the rails have been accurately adjusted, lined and surfaced, the top ballasting shall be completed for a width of 3.15 m and neatly trimmed and boxed flush with the sleepers.

(2) *Sleepers*. Sleepers shall be of creosoted redwood, 2.60 m long and 250 by 125 mm in section laid at 750 mm centres. The faces of the sleepers shall be dressed under each rail to accommodate chairs, etc.

(3) *Rails*. The gauge of the railway track shall be 1435 mm and the rails shall be steel flat bottom (BS113A) and weighing 56 kg/m and supplied in 18 m lengths.

(4) *Fishplates*. The fishplates shall be of the four-hole type and weighing 14 kg/pair. Steel fish bolts and nuts to be 24 mm by 120 mm long, weighing 0.85 kg each.

(5) *Chairs*. The chairs are to be of cast iron, standard variety, weighing 20 kg each, bolted to the sleeper with 3 nr chair bolts 22 mm diameter by 185 mm long with washers 80 mm square, weighing 1 kg each. The keys are to be steel spring keys to railway standard pattern.

(6) *Tracklaying*. The rails shall be accurately laid to line, level, gauge and to the correct radii of the respective curves, with such super-elevation on the outer rail on curves as may be required by the Engineer, and the price for tracklaying must include all these costs.

Metal slips 8 mm thick, shall be inserted in the rail joints to provide expansion spaces and shall be kept in the joints until the rails have been lined and secured.

No closing length of less than 4.50 m shall be used and all cuts in rails shall be square and clean. The prices inserted by the Contractor shall include for all cutting and waste arising out of the tracklaying and the whole of the work shall be carried out in accordance with present-day first-class railway practice.

Note: current practice is normally to use welded rails fastened to prestressed concrete sleepers with Pandrol rail fastenings to give improved performance. The method of measurement is similar to that used for the more traditional track.

			RAILWAY TRACK	EXAMPLE XVIII

RAILWAY TRACK
(150 m length of straight plain track)

<table>
<tr><td></td><td></td><td></td><td colspan="2">Track foundations</td></tr>
<tr><td></td><td>150·00
3·15
0·15</td><td></td><td>Bottom ballast, crushed
stone.</td><td>S110</td></tr>
<tr><td></td><td>150·00
3·15
0·13</td><td></td><td>Top ballast, crushed
stone.</td><td>S120</td></tr>
<tr><td></td><td></td><td></td><td colspan="2">SUPPLY ONLY</td></tr>
<tr><td></td><td></td><td></td><td colspan="2"><u>Rails for plain track</u></td></tr>
<tr><td>2/</td><td>150·00</td><td></td><td>Steel flat bott. rails to
BS113A, mass 56kg/m.</td><td>S424</td></tr>
<tr><td></td><td></td><td></td><td colspan="2"><u>Other materials for
plain track</u></td></tr>
<tr><td></td><td></td><td></td><td colspan="2">750)150·00
200+1</td></tr>
<tr><td>201/</td><td>1</td><td></td><td>Timber sleepers; creosoted
redwood, 250 x 125 x 2600
mm lg.</td><td>S471</td></tr>
<tr><td>201 /</td><td>2</td><td></td><td>Plain chairs; c.i. weighg.
20 kg ea.</td><td>S481</td></tr>
</table>

18.1

Ballast is divided into the two separate items of bottom ballast and top ballast and are both measured in m³ and cover both the supplying and laying. The top ballast item is deemed to include the cost of boxing up, trimming to line and level and tamping after the track has been laid, without the need for specific mention.
No deduction is made from the top ballast for the volume occupied by sleepers (rule M1).

The section reference and mass of the rails is given; the latter over-riding the mass range in the third division as paragraph 5.14 of CESMM3.
This item will be given in tonnes in the Bill.
Supply items are taken first followed by a track laying item in metres.
The number of sleepers is obtained by dividing the length of track by the spacing of sleepers (centre to centre) and adding one to allow for a sleeper at each end of the track.
The supply of sleepers is enumerated stating the type and size and identifying any fittings attached by the supplier (rules A7 and A8).
There are two chairs to each sleeper.
The chair items are deemed to include the required bolts and keys without the need for specific mention (rule C5).
The item description shall include the type of fitting (rule A7).

RAILWAY TRACK   (Contd.)

18)150
      9

| 2/91 | 1 | Prs. of plain fishplates; ms weighg. 14 kg/pr. S484 |

LAYING ONLY

| | 150·00 | Plain track; stl. flat bott. rails to BS113A, mass 56 kg/m, fishplated jts. on timber sleepers. S620 |

SUPPLY ONLY

| | 2 | Turnouts, type T1, dwg. 9TR. S510 |

| | 1 | Diamond crossg., type DCX, dwg. 9TR. S520 |

| | 2 | Sundries, buffer stops, type B1, approx. wt. 2t. S581 |

LAYING ONLY

| | 2 | Turnouts, type T1, dwg. 9TR, len: 25 m, fish pltd. jts. on tbr. sleepers. S624 |

| | 1 | Diamond crossg., type DCX, dwg. 9TR, len. 26m, fish pltd. jts. on timber sleepers. S625 |

| | 2 | Sundries, buffer stops, approx. wt. 2t. S681 |

There are 9 joints to each 150 m length of rail, including the junction with the existing track. There are 2 fishplates to each joint, made up of one plate on each side of the rail.

Fishplates are measured in pairs (rule M6), stating the type. The supply item is deemed to include fixing bolts (rule C5). A linear item for track laying with a brief description of the component parts, as rule A15. Where laid to form a curve an additional linear item would be needed.

Typical items follow for turnouts, diamond crossings and buffers. Items for the supply of turnouts and diamond crossings shall state the type (rule A10). Items for supplying are deemed to include delivery to the site (rule C3).

Item descriptions for supplying buffer stops shall state their approximate weight (rule A12).

Item descriptions for laying turnouts & diamond crossings shall state their type and length (rule A16), while those for buffer stops shall state their approximate weight (rule A18).

18.2

| RAILWAY | TRACK | | (Contd.) |

Takg. up track
Welded track fxd. to conc. sleepers w. Pandrol fastngs, fully dismantled & deposited in Employer's store at Apex Junction.

560·00	Plain track, bullhead rails. S211
2	Bullhead rails, turnouts. S214
2	Sundries, buffer stops, of stl. rail & tbr. sleeper constn., approx. wt. 2t. S281
1	Sundries, wheelstop. S283

Liftg., packg. & slewg.

1	Bullhead rail track, len. 18 m, max. dist. of slew 220 mm & max lift of 120 mm. S310

Typical items follow to illustrate taking up track. The item descriptions for taking up track shall state the amount of dismantling, details of disposal and type of rail, sleeper and joint as given in the heading and item description.

The length of taking up plain track is measured along the centre line of the track and excludes the lengths of turnouts and diamond crossings (rule M3 of class S).

Turnouts and diamond crossings are enumerated.

Item descriptions for buffer stops include their approximate weight and type of construction (rule A4).

Wheelstops form another enumerated item.

Enumerated item for lifting, packing and slewing track, stating the length of track, maximum distance of slew and maximum lift (rule A5), and measured on centre line of track (2 rail). It is deemed to include opening out, packing and boxing in with ballast and insertion of closure rails (rule C2).

18.3

16 Measurement of Sewer and Water Main Renovation and Ancillary Works

Sewers: Renovation Techniques

Renovation is defined in the Water Research Centre (WRc) manual[32] as 'methods by which the performance of a length of sewer is improved by incorporating the original sewer fabric, but excluding maintenance operations such as isolated local repairs and root or silt removal'. WRc has classified renovation techniques under the three broad categories of stabilisation, linings and coating, each of which is now described in outline. CESMM3 class Y incorporates these techniques but also ancillary works involved in the preparation of existing sewers, such as cleaning, removing intrusions, closed-circuit television surveys, plugging and filling laterals and other pipes and local internal repairs.

(1) Stabilisation involves the pointing and chemical grouting of joints.
(2) Linings are of three types: essentially, type I uses the structural capacity of the existing sewer and requires a bond between the lining and the grout, and the grout and the existing structure; type II is designed as a flexible pipe requiring no bond between the lining and the grout or the existing structure, and no long-term strength is assigned to the existing sewer; and type III provides permanent formwork for the grout, and the lining cannot be assumed to contribute to the long-term strength of the sewer. Linings are mainly formed of glass reinforced cement, glass reinforced plastic and resin concrete.
(3) *In situ* gunite forms an effective structural coating.

When considering the renovation option, the five most significant issues are economics, hydraulic capacity, materials considerations, strength and installation aspects. WRc has also given advice on determining the overall hydraulic effect of renovation on the sewer network resulting from changes in cross-sectional area and roughness co-

efficients. With the exception of sliplining in pipe sewers, lining techniques in general do not have a detrimental effect on hydraulic capacity, and many techniques can improve the capacity of brick sewers.
Furthermore, for the leading lining materials sufficient durability data
have been accumulated to justify design life predictions of 50 years. In
most renovation systems, annulus grouting is necessary to ensure
satisfactory performance.

Sewers: Renovation Measurement

Section Y of CESMM3 covers the preparation and renovation of existing sewers, the provision of new manholes to existing sewers and work
to existing manholes. A new item has been inserted in CESMM3 (Y 1 3
0) to cover closed circuit television surveys, as they are becoming used
extensively, normally with a facility for taking in-sewer colour photographs, together with reports coded in accordance with the *Manual of
sewer condition specification*.[33] Specified requirements in Class A
could be used for such work as confirming sewer dimensions and core
sampling. While the Contractor may be given the opportunity to enter
method-related charges for exploratory trenches, temporary access
shafts, pumping, diversionary works and the like.[5]

The principal characteristics of the main sewer are to be stated (rules
A2 and A5), while rules A3 and A4 distinguish between 'man entry' and
'no man entry' sewers where the Engineer dictates the choice of
method.

The preparation of existing sewers under classification Y 1** includes preparatory work such as removal of silt, grease, encrustation
and tree roots, which is carried out prior to sewer renovation. In some
cases, preparation work may form a preliminary contract to help in
establishing the condition of the existing sewers before a realistic
assessment can be made of the stabilisation and renovation work.

Most items of preparation work to existing sewers are enumerated
(removing intrusions, plugging laterals and local internal repairs),
while cleaning and closed-circuit television surveys are measured as
linear items, and filling laterals and other pipes in m^3, with locations
clearly defined as in rule A1. Stabilisation of existing sewers can be
carried out by pointing (in m^2), pipe joint sealing (by number), or
external grouting (grout holes by number and grout injected in m^3).

The measurement of cleaning to existing sewers is classified very
simply in Y 1 1 0, although it will be necessary to identify differing
cleaning requirements and standards for varying locations billed in
accordance with rule A1. The cleaning items are deemed to include
making good damage resulting from the cleaning work but not other-

wise (rule C2). For example, the Contractor cannot be held responsible for sewer damage exposed by the cleaning but not caused by it.

Removing intrusions in existing sewers are classified into three groupings in Y 1 2 1–3. An intrusion is a projection into the bore of the sewer. Artificial intrusions may encompass isolated projecting bricks, projecting rubber O rings and dead services. While laterals comprise any drains or sewers which are connected to the sewer being renovated, which are prepared by cleaning (Y 1 1 0) or by sealing (Y 1 4–5*), with sealing/filling of laterals and other pipes measured in m³. Local internal repairs covered by Y 1 6 1–3 are repairs to the structural fabric of the sewer which are to be carried out from inside the sewer. Typical examples are isolated patch repairs, repairs to bellmouths and Y junctions which are not to be renovated and repairs around laterals.

Stabilisation of existing sewers is carried out by pointing, joint sealing and grouting as Y 2 1–3*. Different types of pointing such as hand pointing and pressure pointing are distinguished by the different locations in accordance with rule A1. External grouting is only measured where it is expressly required to be carried out as a separate operation from annulus grouting in Y 3 6 0 (rule M4), and consists of the grouting of voids outside the existing sewer from inside it other than in the course of annulus grouting (rule D3).

Renovation of existing sewers encompasses a variety of techniques and materials for improving the performance of sewers, each with their distinct advantages and disadvantages, and including striplining (lengths of pipe lining jointed before being moved into permanent positions), *in situ* jointed pipe lining (lengths of pipe linings jointed at permanent positions), segmental lining (circular or non-circular sewer linings made up from pairs of upper and lower segments jointed near their springings), stated proprietary lining, and gunite coating of stated thickness — all measured as linear items.

Where the Contractor is permitted to choose the appropriate renovation technique to be used, the procedure will be described in the preamble as 5.4 of CESMM3. Curved work is defined as *in situ* jointed pipe lining and segmental lining curved to an offset which exceeds 35 mm per metre (rule A9).

Another alternative is the grouting of annular voids, mainly to consolidate the brickwork and seal cracks, and may be used to fill voids which have been formed outside the existing sewer, and is measured in m³.

Laterals may also need realigning or jointing and are inserted as enumerated items, with the item descriptions including the type of lining to which laterals are to be connected and identification of laterals which are to be regraded (rule A10). Lateral items are deemed to include the work involved in connecting to the lining within 1 m from the inside face of the lined sewer (rule C6). Where the grading work involves a longer length of lateral, a separate item is required to cover the work.

The installation of new manholes and the abandonment, removal and replacement of existing manholes are enumerated. Full details of work to existing manholes shall be included in item descriptions (rule A15). Class Y 6** prescribes the rules for the measurement of new manholes and these are identical to those contained in Class K 1** (pipework — manholes) and the associated measurement, definition, coverage and additional description rules. Items for new manholes which replace existing manholes shall be deemed to include breaking out and disposal of existing manholes (rule C11 of Class Y).

Interruptions to work through excessive flows are measured under Clause Y 8** only where a minimum pumping capacity is expressly required and for periods of time during normal working hours, when the flow in the sewer exceeds the installed capacity requested by the Engineer, and the work is interrupted (rule M7). The unit of measurement is the hour, and the Engineer shall prescribe the minimum pumping capacity which he considers adequate. Rules A1 and A2 subdivide the interruption item by location and sewer type and size. The works embrace preparation, stabilisation and renovation of sewers, in addition to work on laterals and manholes.

Water Mains Renovation: Measurement

This is a new section inserted in CESMM3 in parallel with renovation of sewers. Cleaning of water mains to be renovated is measured in metres taken along the centre lines, and shall include the space occupied by fittings and valves and state the nominal bore (Class Y 5 1* and rule M6). Removing intrusions and pipe sample inspections are enumerated, while closed circuit television surveys, cement mortar lining and epoxy lining of existing mains are taken as linear items.

Pipe sample inspections and closed circuit television surveys shall include work carried out either before or after cleaning and lining (rule D5), while items for sample survey inspections shall be deemed to include replacing the length removed by new pipework (rule C7). Item descriptions for linings include the materials, nominal bores and lining thicknesses (rule A11).

Worked Examples

Worked examples follow covering the measurement of sewer and water main renovation and ancillary works.

RODNEY STREET
MANHOLES 8-12
XTG. BK.SEWER, NOM.
SIZE: 1050 X 750 mm,
EGG SHAPED

Prepn. of xtg. sewer

130·00	Cleaning.	Y110
16	Removg. intrusns; lateral, bore ne. 150 mm; clay.	Y121
60·00	CCTV surveys.	Y130
25	Plugging laterals w. grout to Specfn. clause 24.3, bore n.e. 300 mm.	Y141
1.00 0.40 0.30	Fillg. lateral w. grout to Specfn. clause 24.3, int. c.s. dims, 400 x 300 mm U-shaped.	Y152
3	Local intl. reps; area: 0.1 - 0.25 m²	Y162

EXAMPLE XIX

Some typical examples are included for the benefit of readers.

The location of the work is stated to permit identification by reference to Drawings (rule A1 of class Z). Principal dimensions and profiles of sewers are to be given (rule A2). Cleaning item is deemed to include making good resultant damage (rule C2).

Intrusions into bores of existing sewers to be removed prior to renovation (rule D1).

Now becoming a popular technique, giving excellent guidance as to condition.

Plugging laterals not exceeding 300 mm bore taken as an enumerated item, stating the materials.

Filling laterals and other pipes measured in m³ stating the materials used and the stated profile and size.

The area stated in item descriptions is the finished surface area (rule D2) and the item is deemed to include cutting out and repointing (rule C4).

XTG. BK. SEWER, NOM.
SIZE: 1050 x 750 mm,
EGG SHAPED

Stabilization of xtg.
sewers.

15·00 1·80	Pointg. bwk. in c.m. (1:3). Y210	Measured in m³ stating the materials used. No deduction is made for openings or voids not exceeding 0·5 m² in area (rule M3).
9	Sealg. pipe jts., n.e. 300 mm bore, under pressure w. epoxy mortar. Y220	Pipe joint sealing is deemed to include preparation of joints (rule C5).

Extl. groutg.

6	Nr. of holes Y231	External grouting is measured only where grouting is expressly required to be carried out as a separate operation from annulus grouting (rule M4). Where the external grouting is to be carried out through pipe joints, item descriptions for number of holes shall state this (rule A7).
16·00 3·20 0·25	Injection of ct.grt. to Specfn. clause 24.3. Y232	External grouting consists of the grouting of voids outside the existing sewer, other than voids grouted in the course of annulus grouting (rule D3).

Renovatn. of xtg. sewers

Slipling.

66·00	Butt fusion welded (HDPE) conventln. slipling ; design II, min. thickness 4 mm, to egg shaped sewer, nom. size 1050 x 750 mm. Y311	The item description is to include the type of lining, minimum finished internal size and thickness or grade (rule A8).

	Description	Notes
	XTG. BRICK SEWER, NOM. SIZE : 1050 x 750 mm, EGG SHAPED	Note main headings to each section.
	<u>Renovatn. of xtg. sewers</u> <u>Segmental lings.</u>	
<u>47·00</u>	GRP single piece ling., design type III to Specfn. clause 28.3, min. int. c.s. dims. 900 x 600 mm, egg shaped. Y333	Item descriptions of segmental linings shall state the type of lining, its minimum finished internal size and its thickness or grade (rule A8).
<u>52·00</u>	GRC design type I to Specfn. clause 29.4, 15 mm th., min. intl. c.s. dims. 900 x 600 mm, egg shaped. Y334.1	
<u>4·00</u>	GRC, ditto., int. c.s. dims. 1170 x 700 mm, curved to offset of 80 mm/m. Y334.2	The offset shall be stated where the lining is curved to an offset which exceeds 35 mm/m (rule A9). The volume does not include external grouting (rule M5) and consists of the annular voids between new linings and existing sewers and associated work (rule D4).
	<u>Annulus groutg.</u>	
103·00 2·80 0·13	Ct. grt. as Specfn. clause 24.4. Y360	
	<u>Laterals to renovated</u> <u>sewers</u>	Item descriptions shall state the type of lining to which the laterals are to be connected and identifying those laterals which are to be regraded (rule A10), and are deemed to include work involved in connecting to the lining within 1 m from inside face of lined sewer (rule C6).
	<u>Jointing</u>	
<u>4</u>	Bore 150 – 300 mm to HDPE slipling., design II, thickness 4 mm, regraded. Y412	

	XTG. BRICK SEWER, NOM. SIZE: 1050 × 750 mm, EGG SHAPED	Main headings are inserted in capitals to give greater emphasis.

Flap valves

3	Remove xtg; nom. diam. 300 mm. Y421	Removing, replacing and provision of new flap valves are each enumerated, stating the size in each case.
2	Replace xtg., nom. diam. 225 mm. Y422	

New mhs, replacg. xtg. mhs.

2	Brick w. backdrop type 1B; depth 2·5 – 3 m, type CMH3, w.c.i. cover to BS497, ref. B4-22. Y624	The item description follows the procedure prescribed in class K. Item descriptions shall separately identify new manholes which replace existing manholes and they are deemed to include breaking out and disposal of existing manholes (rules A14 and C11).

Xtg. mhs.
Abandonment as Dwg. 14F

1	Depth 2·5 – 3 m, inc. removg. cover slab, demolishg. shaft & backfillg. w. sel. fillg. Y714	Item descriptions for the abandonment or alteration of existing manholes supply the contractor with all the basic information he requires for pricing and are usually supported by relevant drawings which detail the works involved.

Alteratns.

1	Wk. to benchg. & invts. as Dwg. 15E, inc. breakg out, re-haunchg & dealg. w. flows. Y720

Interruptns.

12h	Prepn. of xtg. sewers. Y810	Interruptions are measured only where a minimum pumping capacity is expressly required for periods of time during normal working hours, when the flow in the sewer exceeds the installed pumping capacity and work is interrupted (rule M7).
12h	Stabiliztn of xtg. sewers. Y820	
15h	Renovatn. of xtg. sewers; segmental lings. Y833	

19.4

EXAMPLE XX

VALVES SV5 - 18

XTG. C.I. MAINS, NOM BORE

n.e. 200 - 300 mm

Commence with a suitable heading giving the location of the water mains to be treated. As they all come within the same nominal bore range, they can be arranged under a single heading. The procedure follows closely that adopted for renovating sewers.

220.00   Cleaning.

Y512

Starting with cleaning measured in metres, measured along centre lines of mains, and including lengths occupied by fittings and valves (rule M6 of class Y).

26   Removg. intrusns.

Y522

Removing intrusions and pipe sample inspections are enumerated, and the latter include replacing the length removed by new pipework (rule C7).

8   Pipe sample inspectns.

Y532

140.00   CCTV surveys.

Y542

Closed-circuit television surveys are a linear item.

140.00   Epoxy ling. as specfn. clause 26.8.

Y562

Item descriptions for lining shall state the materials, nominal bores and thicknesses of the lining (rule A11). In this item the additional information is obtained from the specification.

20.1

17 Bill Preparation Processes

Working-up

This chapter is concerned with the final stages leading up to the preparation of bills of quantities for civil engineering work, after the dimensions have been 'taken-off'. The term 'working-up' is applied to all the various operations collectively and can comprise the following processes.

(1) Squaring the dimensions and entering the resultant lengths, areas and volumes in the third or squaring column on the dimensions paper.
(2) Transferring the squared dimensions to the abstract (illustrated in example XXI), where they are written in a recognised order, ready for billing, under the appropriate section headings, and are subsequently totalled and reduced to the recognised units of measurement in readiness for transfer to the bill.
(3) In the bill of quantities, the various items of work making up the project are then listed under appropriate section headings, with descriptions printed in full and quantities given in the recognised units of measurement, as laid down in the *Civil Engineering Standard Method of Measurement (CESMM3)*. The bill also contains rate and price columns for pricing by contractors when tendering for the project.

Billing-direct

The working-up process, which has been used for many decades in quantity surveyors' offices, is very lengthy and tedious, and various ways of shortening this process have been developed. One of the first methods to be introduced was to 'bill direct', by transferring the items direct from the dimension sheet to the bill, thus eliminating the need for an abstract, and so saving both time and money.

The billing-direct system can be used where the number of similar

252

items is not too extensive and the work is not too complex in character. Drainage work is a particular instance where this shorter method can, with advantage, be adopted, particularly when pipework and chamber schedules have been prepared.

With the object of speeding up the working-up process and reducing the staff time involved, further methods using computers occasionally on a national basis, or a 'cut and shuffle' system in the quantity surveying office have been developed. These more recent methods will be described later in this chapter.

Squaring the Dimensions

The term 'squaring the dimensions' refers to the calculation of the numbers, lengths, areas and volumes and their entry in the third or squaring column on the dimensions paper. The following examples illustrate the squaring of typical dimensions on dimensions paper.

When there are timesing figures entered against the item to be squared, it is often simpler to multiply one of the figures in the

Dimensions Notes

7/2/	15.20	212.8	Fwk. fair fin. hor. width: 0.1–0.2 m.	Linear item: Total length is 212.80 m or 212 metres, 800 milli-metres (14 × 15.20 m)
	90.00 10.00	900.0	Fwk. fair fin. vertical.	Square or superficial item: area is 900 m²
	90.00 2.40 1.00	216.0	Placing of mass conc. in bases & grd. slabs, thickness ex. 500 mm.	Cubic item: Volume of concrete is 321 m³. Note method of casting up a series of
	50.00 2.10 1.00	105.0		dimensions relating to the same item with the total entered in the
		321.0		description column and the use of the bracket. Deductions following the main items can be dealt with in a similar manner.

dimension column by the timesing figure before proceeding with the remainder of the calculation. Alternatively, the total obtained by the multiplication of the figures in the dimension column is multiplied by the timesing figure.

The squaring must be checked by another person to eliminate any possibility of errors occurring. All squared dimensions and waste calculations should be ticked in coloured ink or pencil on checking and any alterations made in a similar manner. Amended figures need a further check. Where, as is frequently the case, calculating machines are used for squaring purposes a check should still be made.

Abstracting

An example showing typical completed abstract sheets is given in example XXI, later in this chapter, and the items will subsequently be produced in bill form in example XXII. The abstract in example XXI covers the dimensions for the stone-faced sea wall taken-off in example VIII (chapter 8), where the dimensions have been squared in readiness for abstracting. As each item is transferred to the abstract the description of the appropriate dimension item is crossed through with a vertical line on the dimension sheet, with short horizontal lines at each end of the vertical line, so that there shall be no doubt as to what has been transferred.

The abstract sheets are ruled with a series of vertical lines spaced about 25 mm apart and are usually of A3 width.

Each abstract sheet is headed with the project reference, sheet number and section of the work, and possibly the sub-section of the work, to which the abstracted dimensions refer. The section headings normally follow those given in the *Civil Engineering Standard Method of Measurement* (CESMM3) and are usually produced in the same order.

Entries in the abstract should be well spaced and it is necessary for the worker-up to look through the dimension sheets, before he starts abstracting, in order to determine, as closely as possible, how many abstract sheets will be required. They should not be any closer than the entries in example XXI.

The items will be entered in the abstract in the same order as they will appear in the bill, as far as is practicable, since the primary function of the abstract is to classify and group the various items preparatory to billing, and to reduce the dimensions to the recognised units of measurement. Descriptions are usually spread over two columns with the appropriate dimension(s) in the first column and any deductions in the second column. The total quantity of each item is reduced to the recognised unit of measurement such as kilogrammes to tonnes.

It is good practice to precede each description in the abstract with the prefix C, S, L or Nr, denoting that the item is cubic, square, linear or enumerated to reduce the risk of errors arising with regard to units or quantities.

As to the order of items in each section or sub-section of the abstract, the usual practice is to adopt the order of cubic, square, linear and finally enumerated items, with labour items preceding labour and materials, smaller items preceding larger ones and cheaper items preceding the more expensive in each group, but at the same time also having regard to the sequence of items in CESMM3 including the order of code numbers.

Where it is necessary to abstract a number of similar items but of different sizes, the best procedure is to group these items under a single heading with each size entered in a separate column, as shown in the following example.

Cast (spun) iron s & s pipes to BS 1211 (class B) w. caulked lead jts. in trenches, depth ne 1.5 m

150 mm dia.	225 mm dia.	300 mm dia.	375 mm dia.
154.00 (6)	104.00 (7)	226.00 (10)	204.00 (11)
186.00 (9)	192.00 (8)	176.00 (11)	142.00 (12)
218.00 (10)	184.00 (9)		

The number entered in brackets after the dimension represents the page number of the dimension sheet from which the dimension has been extracted, for ease of reference.

All squaring and abstracting work and the transfer of the abstract items to the bill must be checked by a second person to verify their accuracy.

Billing

Example XXII, given later in this chapter, incorporates the billed items for the stone-faced sea wall, based on the entries in the abstract forming example XXI. As each item is transferred to the bill it is crossed through on the abstract to prevent any possibility of errors occurring during the transfer stage.

The order of billed items will be the same as in the abstract, as far as is practicable, and they will be grouped under suitable section headings. There may be a number of preamble clauses at the head of each section relating to financial aspects of the work in the section concerned and giving guidance to the Contractor in his pricing of the items.

Each item in the bill is indexed, usually by the numbering of items in the first column possibly incorporating CESMM3 coding. It will be noticed that all words in the billed descriptions are inserted in full without any abbreviations and this procedure should always be followed to avoid any possible confusion.

Provision is usually made for the total sum on each page of the bill relating to a given section of work to be transferred to a collection at the end of the section. The total of each of the collections is transferred to a Grand Summary, the total of which will constitute the tender total. This procedure is preferable to carrying forward the total from one page to another in each section, since the subsequent rectification of errors in pricing may necessitate alterations to a considerable number of pages.

Billed descriptions should conform to the requirements of CESMM3, follow in a logical sequence and be concise, yet must not, at the same time, omit any matters which will be needed by the Contractor if he is to be able realistically to assess the price for each item.

The first bill is likely to cover General Items incorporating Specified Requirements and Method-related Charges, where the Contractor can price items that are not proportional to the quantities of Permanent Works.

Recent Developments in Bill Preparation

General Introduction

New measurement and processing techniques have been introduced in recent years and they are now being used to an increasing extent, since they are resulting in a speeding up of working-up operations and a reduction in the overall cost of preparing bills of quantities.

Over the years many quantity surveyors and engineers have experimented with a number of systems designed to eliminate part of the working-up process. These systems include the elimination of the abstract by billing direct as described earlier in this chapter, taking-off direct on to abstract sheets and using full descriptions in the abstract to permit the abstract to be edited as a draft bill. It was, however, generally found that each of these systems could only function satisfactorily under a certain set of conditions and were not, therefore, of universal application.

'Cut and Shuffle'

The system of 'cut and shuffle' was developed in the early 1960s and by the late 1970s was probably the most widely used method of entering dimensions and descriptions. It has been aptly described as a rationalised traditional approach. Unlike abstracting and billing there is no universally accepted format and many different paper rulings and methods of implementation are used in different offices.

However, the following criteria apply to most systems:

(1) Dimensions paper is subdivided into four or five separate sections which can subsequently be split into individual sections.
(2) Only one description with its associated dimensions is written on each section.
(3) Dimension sheets are subsequently split into separate slips and sorted into bill work sections and eventually into bill order.
(4) Following the intermediate processes of calculation and editing, the slips form the draft for producing the final bill of quantities.

The cut and shuffle method is designed to eliminate the preparation and checking of the abstract and the draft bill. Hence there is only one major written operation, namely taking-off, compared with the three entailed with abstracting and billing.

One method of carrying out the technique is now described.

(1) Taking-off is carried out on A4 sheets of dimensions paper, ruled vertically into four columns, and thus accommodating four items per sheet. Dimensions are entered on one side only of each sheet and each column is generally stamped with the project reference and numbered consecutively. 'Ditto.' items must include a reference to the column number of the main item, where full particulars can be found.
(2) As sections of the taking-off are completed, the side casts are checked and repeat dimensions calculated.
(3) When the taking-off is complete, each column is marked with the taking-off section number, work section reference and column number. A copy of each dimension sheet is obtained, generally either by using NCR (no carbon required) paper or by photocopying. However, some systems operate without the need to produce a copy.
(4) The taker-off retains the copy and the original sheet is cut into four slips, each containing one item. Some quantity surveyors use sheets that are already perforated.
(5) The slips are shuffled or sorted into sections, such as Earthworks,

In situ Concrete, Concrete Ancillaries and Precast Concrete. Similar items are collected together and the whole of the slips placed, as near as possible, in bill order.

(6) When all the slips for an individual work section have been sorted, they are edited to form the draft bill, with further slips being inserted as necessary to provide headings, collections and other relevant items. The correct unit is entered on the 'parent' or primary item slip and the 'children' or repeat item slips are marked 'a.b.' (as before). As each section is edited it is passed to a calculator operator for squaring.

(7) The calculator operator squares, casts, reduces and inserts the reduced quantity on the parent item slip. This operation is double checked.

(8) Parent and children slips are separated. The parent slips form the draft bill and are ready for processing.

(9) Any further checks on the draft bill are then carried out and final copies made and duplicated.

(10) The children slips are then replaced to provide an abstract in bill order for reference purposes during the post-contract period.

Microcomputers and Bill Production

Many makes of microcomputer came on to the market in the 1980s at constantly reducing prices. They generally consist of a combined keyboard and monitor (visual display unit or screen), one or two floppy disk (disc) drives and a printer of selected speed and quality. The floppy disks are used extensively as they are relatively cheap and possess quite fast data transfer speeds. However, the hard disk systems provide increased storage capacity and higher operating speeds. The main disadvantage of microcomputers stems from the incompatibility of much of the equipment. The risk of loss or damage is easily overcome by making duplicate copies of all disks.

An increasing use is being made of microcomputers for the preparation of bills of quantities, resulting from the reducing cost of hardware (equipment) and the improved range and efficiency of software (programs). This has culminated in the provision of a more efficient, improved and faster service. The newer models enable more data to be held on the computer with greater ease of access. An increasing number of software packages (programs) are available and these permit the production of bills in different formats.

Computer-aided bill production systems provide the facility to check accuracy, but care is needed in the coding of dimensions and entry of data. Modern computerised billing systems can, however, print out errors in the form of tables. The coding can be double checked, although a random check may be considered adequate. The need to engage outside agencies for computerised bill production has been largely eliminated.

Hence the use of computer-aided bill of quantities production packages eliminates the reducing, abstracting and billing operations by converting coded dimension sheets into bills of quantities. Codes are frequently based on the CESMM3 reference codes to form a standard library of descriptions. Items not covered by the standard library are termed rogue items but these are unlikely to be very extensive. The rogue items are suitably coded and entered into either the standard library or the particular project library. The computer normally prints a master copy of the bill of quantities which can be photocopied on to ruled paper to give a high standard of presentation.

Data input systems vary and some offices prefer every taker-off to have a work station and to input his own dimensions and descriptions and/or codes as he proceeds. An alternative is for the data to be collected centrally and to be input by a machine operator, thus permitting the taker-off to avoid the coding if he wishes. It is mainly a question of finance, as to whether it is cheaper for the taker-off to spend a little longer and look up his own codes or whether it is better for a lower-paid member of the staff to do the work. Local circumstances will normally provide the answer, and the size of office and type of workload will have an influence.

There are various types of data input system available. Thus the input method can incorporate traditionally prepared dimensions, suitably coded, or the organisation can use direct keyboard entry with automatic squaring or a fully integrated digitiser.

Range of Computer Programs

Microcomputers can be used to advantage in many activities associated with civil engineering projects and the following selection gives an indication of their wide range and scope, stemming from their extensive storage capacity, and ease of retrieval of data and monitoring of progress.

(1) bills of quantities production
(2) automatic measurement of some civil engineering works
(3) materials scheduling, possibly linked with computer-aided design (CAD)
(4) steel reinforcement calculations
(5) earthwork calculations (cut and fill)
(6) specification production
(7) feasibility studies and cost control

 (8) cost reporting
 (9) estimating and tendering
(10) tender analysis
(11) budgetary controls
(12) valuations
(13) formula price adjustment
(14) variations and final accounts
(15) fee management
(16) quotations and enquiries
(17) cash flow forecasting
(18) capital programming
(19) project planning and control
(20) progress statements
(21) resource analysis
(22) sub-contractors' payments
(23) maintenance scheduling
(24) plant and equipment scheduling.

ICEPAC

The leading software system for civil engineering bill preparation and contract administration is the ICEPAC (Telford) System. This incorporates the CESMM3 Standard Library, operates on a wide range of hardware, has direct access to the CESMM3 Price Database, links to CAD and the Moss Drainage System, and has digitiser input and many other facilities.

CESMM3 Library of Item Descriptions

The CESMM3 Library of Item Descriptions[34] interprets the relevant rules and allows the user to select appropriate information for addition to the basic description derived from CESMM3. The aim is to provide sufficient information whereby a user can build up complete item descriptions in 90 per cent of cases by direct reference to the library.

The library will enhance the manual preparation of civil engineering bills using traditional methods by eliminating the need to refer regularly to other sources to produce complete item descriptions. The library has been compiled with computer usage in mind and hence its adaptation to the needs of computer systems is relatively simple. The library is available in several commercially available software packages, of which the leading one is ICEPAC as previously described.

For layout reasons, and for ease of reading, this page has intentionally been left blank.

ABSTRACT OF DIMENSIONS OBTAINED FROM EXAMPLE VIII (Chapter 8)

EARTHWORKS   1

Excavn.
^{C}/Gen. excavn., max. depth:
0·25 − 0·5 m.

E422

23·4 (7)

= 23 m³

Excavn. Ancillaries
^{S}/ Prepn. of excvtd. surfs.

E522

240·0 (3)

= 240 m²

^{C}/Disposal of excvtd. mat.

E532

^{9}/Gen. excavn., max. depth: 1 − 2 m.

E424

1176·9 (2)   _Ddt._
23·4 (7)   253·5 (2)

91·1 (1)

1200·3
253·5

946·8

= 91 m³

= 947 m³

^{9}/Fillg. to structures.

E613

^{9}/Gen. excavn., max. depth: 5 − 10 m; Commg. Surf. hwl.

E426

253·5 (2)

= 254 m³

1085·8 (1)

= 1086 m³

Note : The deductions are crossed through after transfer.

262

IN SITU CONCRETE

C/ Proven. of conc. designed mix, grade C10, ct to BS 12, 40 mm agg. to BS 882. F224		C/ Placg. of conc. mass walls thickness: ex. 500 mm; backg. to masonry below hwl. F544.1		Conc. Ancillaries S/Fwk. ro. fin. slopg; width: 0.2-0.4 m; below hwl. G123.1
1012.6 (3) 89.1 (4) 1101.7 145.3 956.4 = 956 m³	Ddt. 128.5 (5) 16.8 (6) 145.3	1012.6 (3) 128.5 884.1 = 884 m³	Ddt. 128.5 (5)	21.0 (4) = 21 m²
				S/Fwk. ro. fin. vert. width: 0.4 - 1.22 m; above hwl. G144.1
		C/ Ditto. above hwl. F544.2		66.0 (5) = 66 m²
		89.1 (4) 16.8 72.3 = 72 m³	Ddt. 16.8 (6)	
Note: C denotes cubic items, S superficial ones, L linear ones and Nr are enumerated items. Numbers in brackets denote dimension sheet page numbers. In a straightforward project like this with a restricted number of items and where the quantities can be totalled on the dimension sheets, it is quite feasible to omit the abstract and transfer the quantities direct from the dimension sheets to the bill.				S/Fwk. ro. fin. vert; below hwl. G145.1
				420.0 (4) = 420 m²

Interlockg. steel piles type 2N, setn. modulus 1150 cm³/m, grade 43A to BS4360. ⁹/Driven area. P832					
225·0 (3) = 225 m²					
		⁹/Ash base, depth: 75 mm. R713			
		180·0 (7) = 180 m²			
⁹/Area of piles of len: n.e. 14 m; treated w. 2 cts. bit. paint. P833		⁹/Red precast conc. flgs. to BS7263 type D; thickness: 50 mm. R782			
225·0 (3) = 225 m²		180·0 (7) =180 m²			

264

MASONRY   4

Ashlar masonry, Portland stone, flush ptd. w. mortar type M3. S/ Vert. st. wall; thickness: 300 mm, fair faced b.s. U731 27.0 (6) = <u>27 m²</u>	4/ Copg. 1125 x 600 mm rdd. w. sinkgs. as Dwg. 10. U771·1 60.0 (6) = <u>60 m</u>	
S/ Battered fcg. to conc; thickness: nom. 400 mm above hwl. U736·1 42.0 (5) = <u>42 m²</u>	4/ Copg. 450 x 225 mm 2ce wethd. & 2ce thro. as Dwg. 10. U771·2 60.0 (6) = <u>60 m</u>	4/ Dpc; width: 300 mm, 2 cos of slates, ld. bkg. jt. in mortar type M3. U782 60.0 (7) = <u>60 m</u>
S/ Ditto. thickness: nom. 600 & 400 mm in alt. cos. av. 514 mm, below hwl. U746·1 252.0(5) = <u>252 m²</u>	L/ Plinth 525 x 600 mm 2ce splyd. as Dwg. 10. U777·1 60.0 (6) = <u>60 m</u>	

Example XXII — Bill of Quantities for Stone-faced Sea Wall
(prepared from Abstract in example XXI)

Number	Item Description	Unit	Quantity	Rate	Amount £	p
	EARTHWORKS					
	General Excavation					
E422	Maximum depth 0.25–0.5 m.	m^3	23			
E424	Maximum depth: 1–2 m.	m^3	91			
E426	Maximum depth: 5–10 m; Commencing Surface: high water level.	m^3	1086			
	Excavation Ancillaries					
E522	Preparation of excavated surfaces.	m^2	240			
E532	Disposal of excavated material.	m^3	947			
	Filling					
E613	Filling to structures.	m^3	254			
			Page total			

Bill of Quantities (contd.) Stone-faced Sea Wall

Number	Item Description	Unit	Quantity	Rate	Amount £	p
F224	IN SITU *CONCRETE* *Provision of Concrete* Designed mix, grade C10, cement to BS 12, 40 mm aggregate to BS 882.	m³	956			
F544.1 F544.2	*Placing of Concrete mass walls thickness: exceeding 500 mm; backing to masonry* Below high water level. Above high water level.	m³ m³	884 72			
G123.1 G144.1 G145.1	*CONCRETE ANCILLARIES* *Formwork rough finish* Sloping, width: 0.2–0.4 m; below high water level. Vertical, width: 0.4–1.22 m; above high water level. Vertical; below high water level.	m² m² m²	21 66 420			
			Page total			

Bill of Quantities (contd.) Stone-faced Sea Wall

Number	Item Description	Unit	Quantity	Rate	Amount £	p
	PILES *Interlocking steel piles type 2N, section modulus 1150 cm³/m, grade 43A to BS 4360.*					
P832	Driven area.	m²	225			
P833	Area of piles of length: not exceeding 14 m; treated with two coats of bitumen paint.	m²	225 Total			
	PAVINGS *Light duty pavement*					
R713	Ash base, depth: 75 mm.	m²	180			
R782	Red precast concrete flags to BS 7263 type D; thickness: 50 mm.	m²	180			
			Total			
	Note: Two different classes have been inserted on the same page because of the small number of items involved. Hence it is necessary to have two totals for transfer to the Grand Summary.					
				Page total		

Bill of Quantities (contd.) Stone-faced Sea Wall

Number	Item Description	Unit	Quantity	Rate	Amount	
					£	p
	MASONRY *Ashlar masonry Portland stone flush pointed with mortar type M3*					
U731	Vertical stone wall; thickness: 300 mm, fair faced both sides.	m²	27			
U736.1	Battered facing to concrete; thickness; nominal 400 mm, above high water level.	m²	42			
U746.1	Ditto., thickness: nominal 600 and 400 mm, in alternate courses, average 514 mm, below high water level.	m²	252			
U771.1	Coping 1125 × 600 mm rounded with sinkings as Drawing 10.	m	60			
U771.2	Coping 450 × 225 mm, twice weathered and twice throated as Drawing 10.	m	60			
U777.1	Plinth 525 × 600 mm, twice splayed as Drawing 10.	m	60			
U782	Damp-proof course; width: 300 mm, two courses of slates laid breaking joint in mortar type M3.	m	60			
			Page total			

References

1. Institution of Civil Engineers and Federation of Civil Engineering Contractors. *Civil Engineering Standard Method of Measurement, Third Edition: CESMM3.* Telford (1991).
2. McCaffer and Baldwin. *Estimating and Tendering for Civil Engineering Works.* BSP (1991).
3. Davis, Langdon and Everest. *Spon's Civil Engineering and Highway Works Price Book* (1992).
4. *Wessex Database for Civil Engineering.* Wessex Electronic Publishing (1990).
5. M. Barnes. *CESMM3 Handbook.* Telford (1992).
6. N.M.L. Barnes and P.A. Thompson. *Civil Engineering Bills of Quantities.* CIRIA Report 34. Construction Industry Research and Information Association (1971).
7. I.H. Seeley. *Public Works Engineering.* Macmillan (1992).
8. Royal Institution of Chartered Surveyors and Building Employers Confederation (RICS and BEC). *Standard Method of Measurement of Building Works: SMM7* (1988).
9. Co-ordinating Committee for Project Information. *Co-ordinated Project Information for Building Works: a guide with examples* (1987).
10. I.H. Seeley. *Building Quantities Explained*, Fourth Edition. Macmillan (1988).
11. I.H. Seeley. *Advanced Building Measurement*, Second Edition. Macmillan (1989).
12. *Hudson's Building and Engineering Contracts.* Sweet and Maxwell (1992).
13. Institution of Civil Engineers, Association of Consulting Engineers and Federation of Civil Engineering Contractors. *ICE Conditions of Contract*, Sixth Edition (January 1991).
14. Institution of Civil Engineers. *Civil Engineering Procedure.* Telford (1986).
15. I.H. Seeley. *Civil Engineering Contract Administration and Control.* Macmillan (1986).

16. C.K. Haswell and D.S. de Silva. *Civil Engineering Contracts: Practice and Procedure.* Butterworths (1989).
17. R.J. Marks, R.J.E. Marks and R.E. Jackson. *Aspects of Civil Engineering Contract Procedure.* Pergamon (1985).
18. I.H. Seeley. *Quantity Surveying Practice.* Macmillan (1984).
19. The Aqua Group. *Tenders and Contracts for Building.* BSP (1990).
20. Institution of Civil Engineers, Association of Consulting Engineers and Federation of Civil Engineering Contractors. *ICE Conditions of Contract for Minor Works* (1988).
21. Fédération Internationale des Ingénieurs-Conseils and the Fédération Internationale Européenne de la Construction. *The Conditions of Contract (International) for Works of Civil Engineering Construction* (1987).
22. Institution of Civil Engineers. *The New Engineering Contract.* Telford (1991).
23. Joint Contracts Tribunal for the Standard Form of Building Contract. *Standard Form of Building Contract* (1980).
24. *General Conditions of Government Contracts for Building and Civil Engineering Works.* GC/Wks/1. HMSO (December 1989).
25. Institute of Quantity Surveyors, Civil Engineering Committee. *A Report on Procedure in connection with the ICE Conditions of Contract* (1973).
26. Royal Institution of Chartered Surveyors, Quantity Surveyors Civil Engineering Working Party. *The ICE Conditions of Contract* (1973).
27. Federation of Civil Engineering Contractors. *Schedules of Dayworks carried out incidental to Contract Work* (1983) and subsequent amendments.
28. Association of Surveyors in Civil Engineering. *The Fifth Edition Explained: Notes for Guidance on the ICE Conditions of Contract — Fifth Edition, revised 1979* (1979).
29. Department of the Environment. *Monthly Bulletin of Construction Indices (Civil Engineering Works).* HMSO.
30. I.H. Seeley. *Civil Engineering Specification.* Macmillan (1976).
31. Ministry of Transport. *Specification for Highway Works.* HMSO (1986).
32. Water Authorities Association and Water Research Centre. *Sewerage Rehabilitation Manual.* WRc (1986).
33. National Water Council and DoE Standing Technical Committee. *Manual of Sewer Condition Classification.* Report 24. NWc (1980).
34. R.E.N. McGill. *CESMM3 Library of Item Descriptions.* Telford (1991).

Appendix I—Abbreviations

a.b. *as before*
a.b.d. *as before described*
additnl. *additional*
adj. *adjoining*
a.f. *after fixing*
agg. *aggregate*
alt. *alternate*
ancills. *ancillaries*
appd. *approved*
ard. *around*
art. *artificial*
asp. *asphalt*
attchd. *attached*
av. *average*
A.V. *air valve*

backg. *backing*
battg. or batterg. *battering*
bd. *board*
bdg. *boarding*
bearg. *bearing*
beddg. *bedding*
bellmth. *bellmouth*
benchg. *benching*
b.f. *before fixing*
b.i. *build in*
bit. *bitumen* or *bitumastic*
bk. *brick*
bkg. *breaking*
bldg. *building*
b.o.e. *brick on end*
borg. *boring*
bott. *bottom*

b. & p. *bed and point*
br. *branch*
brr. *bearer*
b.s. *both sides*
BS *British Standard*
bwk. *brickwork*

cal. plumb. *calcium plumbate*
calkg. *caulking*
cap. *capacity*
cast. *casement*
cat. *catalogue*
ccs. *centres*
c. & f. *cut and fit*
chan. *channel*
chbr. *chamber*
chfd. *chamfered*
chn-lk. *chain link*
chy. *chimney*
c.i. *cast iron*
circ. *circular*
circum. *circumferential*
c.m. *cement mortar*
commg. or commncg.
   commencing
comp. *composite*
compactn. *compaction*
conc. *concrete*
conn. *connection*
constn. *construction*
c.o.p. *circular on plan*
copg. *coping*
cos. *course(s)*

covg. *covering*
c. & p. *cut and pin*
Cr. *Contractor*
c.s. *cross-section*
csa *cross-sectional area*
csg *clear sheet glass*
ct. *cement* or *coat*
cu *cubic*

ddt. *deduct*
deckg. *decking*
delvd. *delivered*
dep. *deposit*
dia. or diam. *diameter*
diag. *diagonally*
dim. or dimng. *diminishing*
dist. *distance*
do. *ditto. (that which has been said before)*
dp. *deep*
d.p.c. *damp-proof course*
dr. *door*
drvg. *driving*
DTp *Department of Transport*
dwg. *drawing*

ea. *each*
embankt. *embankment*
eng. *engineering*
Eng. *English*
Engr. *Engineer*
ent. *entrance*
e.o. *extra over*
ex. *exceeding* or *extra*
exc. *excavate*
excavn. *excavation*
ext. *external* or *externally*

facewk. *facework*
fcg. or facg. *facing*
fdn. *foundation*
f.f. *fair face*
fillg. *filling*
fin. *finish*

fittg. *fitting*
f.l. *floor level*
Flem. *Flemish*
flex. *flexible*
floatg. *floating*
flr. *floor*
F.O. *fix only*
follg. *following*
form. *formation*
fr. *frame*
frd. *framed*
frg. *framing*
frt. *front*
ftg. *footing*
fwd. *forward*
fwk. *formwork*
fxd. *fixed*
fxg. *fixing*

galvd. *galvanised*
gen. *general*
g.i. *galvanised iron*
g.l. *ground level*
glzg. *glazing*
g.m. *gauged mortar*
grano. *granolithic*
grd. *ground*
grdg. *grading*
greenht. *greenheart*
grtd. *grouted*
g.s. *general surfaces*
gtg. *grating*
gth. *girth*

H.A. *highway authority*
ha *hectare*
h.b. *half brick*
h.c. *hardcore*
hi. *high*
holl. *hollowed*
hor. *horizontal*
h.r. *half-round*
ht. *height*
hwd. *hardwood*

h.w.l. *high water level*
H.W.O.S.T. *high water of spring
  tides*

inc. *including*
int. *internal* or *internally*
intl. *internal*
invt. *invert*
irreg. *irregular*

jt. *joint*
jtd. *jointed*
junctn. *junction*

kg. *kilogramme(s)*
km *kilometre(s)*
k.p. & s. *knot, prime and stop*

l. *labour*
la. *large*
L.A. *local authority*
layg. *laying*
len. *length*
lev. *level*
lg. *long*
lin. *linear*
ling. *lining*
l.m. *lime mortar*
long. *longitudinal*
l.w.l. *low water level*

m *metre(s)*
matl. or mat. *material*
max. *maximum*
mech. *mechanically*
med. *medium*
memb. *membrane*
mesd. *measured*
met. *metal*
m.g. *make good*
m.h. *manhole*
min. *minimum*
mm *millimetre(s)*
mo. *mortar*

mors. *mortice*
m.s. *mild steel*
m/s. *measured separately*

nat. *natural*
n.e. *not exceeding*
nec. *necessary*
nom. *nominal*
nr. *number*
n. & w. *nut and washer*
n.w. *narrow widths*

o/a *overall*
O.D. *Ordnance Datum*
o'hg. *overhang*
opg. *opening*
ord. *ordinary*
orig. *original*
oslg. *oversailing*
ov'll *overall (alternative to o/a)*

patt. *pattern*
pavg. *paving*
p.c. *prime cost*
P.ct. *Portland cement*
perm. *permanent*
p.hse. *pumphouse*
pilg. *piling*
p.m. *purpose made*
ppt. *parapet*
pr. *pair*
prepd. *prepared*
prepn. *preparation*
proj. *projection*
provsnl. *provisional*
psn. *position*
p.s. *pressed steel*
P.st. *Portland stone*
pt. *paint*
ptd. *pointed*
ptg. *pointing*
ptn. *partition*
pumpg. *pumping*

qual. *quality*

rad. *radius*
rakg. *raking*
r.c. or r. conc. *reinforced concrete*
rd. *road*
rdd. *rounded*
reb. *rebate*
rec. *receive*
red. *reduced*
ref. *reference*
reinfd. *reinforced*
reinft. *reinforcement*
reqd. *required*
ret. *retaining*
retd. *retained* or *returned*
retn. *return*
r. & g. *rubbed and gauged*
r.h. *rivet head*
r.l. *red lead*
rly. *railway*
ro. *rough*
r.s. *rolled steel*
r.s.j. *rolled steel joist*
r.w. *rainwater*

scrd. *screwed*
sec. or sectn. *section*
seedg. *seeding*
settg. *setting*
sk. *sunk*
s.l. *short length*
sli. *slight*
slopg. *sloping*
sm. *small*
smth. *smooth*
soc. *socket*
soff. *soffit*
spec. *specification*
specd. or specfd. *specified*
spld. *splayed*
sq. *square*
s.q. *small quantities*

s. & s. *spigot and socket*
st. *stone* or *straight*
stan. *stanchion*
stand. *standard*
stl. *steel*
stlwk. *steelwork*
strt. *straight*
struct. *structure*
surf. *surface*
surrd. *surround*
susp. *suspended*
S.V. *sluice valve*
S.W. *surface water*
swd. *softwood*

t *tonne*
tankg. *tanking*
tapd. *tapered*
tarmac. *tarmacadam*
tbr. *timber*
tempy. *temporary*
t. & g. *tongued and grooved*
th. *thick*
thro. *through* or *throated*
timbg. or timberg. *timbering*
tr. *trench*
trimmg. *trimming*
trowld. *trowelled*

UB *Universal beam*
UC *Universal column*
u/c *undercoat*
u/s *underside*

vert. *vertical*
vol. *volume*

w. *with*
W.A. *water authority*
walg. *waling*
wd. *wood*
wdw. *window*
wethd. *weathered*
w.i. *wrought iron*

wk. *work*
W.O. *wash-out*
workg. *working*
w.p. *waterproof*
wrot. *wrought*
wt. *weight*

xtg. *existing*

Y.st. *York stone*

Note: The abbreviation CESMM3 has been used extensively throughout this book and refers to the *Civil Engineering Standard Method of Measurement, Third Edition*.

Appendix II—Mensuration Formulae

Figure	Area
Square	(side)^2
Rectangle	length $\times$ breadth
Triangle	$\frac{1}{2}$ $\times$ base $\times$ height or $\sqrt{[s(s-a)(s-b)(s-c)]}$ where $s = \frac{1}{2}$ $\times$ sum of the three sides and a, b and c are the lengths of the three sides
Hexagon	2.6 $\times$ (side)^2
Octagon	4.83 $\times$ (side)^2
Trapezoid	height $\times$ $\frac{1}{2}$ (base + top)

Circle	(22/7) $\times$ radius^2 or (22/7) $\times$ $\frac{1}{4}$ diameter^2
	    (πr^2) ($\pi D^2/4$)
	circumference = 2 $\times$ (22/7) $\times$ radius or
	                ($2\pi r$)
	    (22/7) $\times$ diameter
	        (πD)

Sector of circle	$\frac{1}{2}$ length of arc $\times$ radius
Segment of circle	area of sector $-$ area of triangle

Figure	*Volume*	*Surface Area*
Prism	area of base × height	circumference of base × height
Cube	(side)^3	6 × (side)^2
Cylinder	(22/7) × radius^2 × length ($\pi r^2 h$)	2× (22/7) × radius × (length + radius) [$2\pi r(h + r)$]
Sphere	(4/3) × (22/7) × radius^3 (4/3πr^3)	4 × (22/7) × radius^2 (4πr^2)
Segment of sphere	(22/7) × (height/6) × (3 radius^2 + height^2) [(πh/6) × (3r^2 + h^2)]	curved surface = 2 × (22/7) × radius × height (h) (2πrh)
Pyramid	⅓ area of base × height	½ circumference of base × slant height
Cone	⅓ × (22/7) × radius^2 × height (⅓ $\pi r^2 h$)	(22/7) × radius × slant height (l) (πrl)
Frustum of pyramid	⅓ height [$A + B + \sqrt{(AB)}$] where A is area of large end and B is area of small end.	½ mean circumference × slant height
Frustum of cone	(22/7) × ⅓ height ($R^2 + r^2$ + Rr) where R is radius of large end and r is radius of small end. [⅓ $\pi h(R^2 + r^2 + Rr)$]	(22/7) × slant height ($R + r$) [$\pi l(R + r)$] where l is slant height

For Simpson's rule and prismoidal formula see chapter 6.

Appendix III—Metric Conversion Table

Length   1 in. = 25.44 mm [approximately 25 mm, then
                                   $(\text{mm}/100) \times 4 = \text{in.}$]
               1 ft   = 304.8 mm (approximately 300 mm)
               1 yd = 0.914 m (approximately 910 mm)
               1 mile = 1.609 km (approximately 1⅗km)
               1 m   = 3.281 ft = 1.094 yd (approximately 1.1 yd)
                                   (10 m = 11 yd approximately)
               1 km = 0.621 mile (⅝ mile approximately)

Area   1 ft²   = 0.093 m²
               1 yd² = 0.836 m²
               1 acre = 0.405 ha [1 ha (hectare) = 10 000 m²]
               1 mile² = 2.590 km²
               1 m²   = 10.764 ft² = 1.196 yd²
                                   (approximately 1.2 yd²)
               1 ha   = 2.471 acres (approximately 2½ acres)
               1 km² = 0.386 mile²

Volume   1 ft³   = 0.028 m³
               1 yd³ = 0.765 m³
               1 m³   = 35.315 ft³ = 1.308 yd³
                                   (approximately 1.3 yd³)
               1 ft³   = 28.32 litres (1000 litres = 1 m³)
               1 gal   = 4.546 litres
               1 litre = 0.220 gal (approximately 4½ litres to the
                      gallon)

Mass   1 lb   = 0.454 kg (kilogram)
               1 cwt = 50.80 kg (approximately 50 kg)
               1 ton = 1.016 t [1 tonne = 1000 kg = 0.984 ton]
               1 kg   = 2.205 lb (approximately 2⅕ lb)

Density

1 lb/ft³ = 16.019 kg/m³
1 kg/m³ = 0.062 lb/ft³

Velocity

1 ft/s   = 0.305 m/s
1 mile/h   = 1.609 km/h

Energy

1 therm   = 105.506 MJ
1 Btu   = 1.055 kJ

Thermal conductivity

1 Btu/ft² h °F = 5.678 W/m² °C

Temperature

x °F   = $\frac{5}{9}(x - 32)$ °C
x °C   = $\frac{9}{5}(x + 32)$ °F
0 °C   = 32 °F (freezing)
5 °C   = 41 °F
10 °C   = 50 °F (rather cold)
15 °C   = 59 °F
20 °C   = 68 °F (quite warm)
25 °C   = 77 °F
30 °C   = 86 °F (very hot)

Pressure

1 lbf/in.²   = 0.0069 N/mm² = 6894.8 N/m²
                  (1 MN/m² = 1 N/mm²)
1 lbf/ft²   = 47.88 N/m²
1 tonf/in.²   = 15.44 MN/m²
1 tonf/ft²   = 107. 3 kN/m²

For speedy but approximate conversions

1 lbf/ft² $= \dfrac{kN/m^2}{20}$ hence 40 lbf/ft² = 2 kN/m²

and tonf/ft² = kN/m² × 10, hence, 2 tonf/ft² = 20 kN/m²

Floor loadings office floors — general usage: 50 lbf/ft² = 2.50 kN/m²
 office floors — data-processing equipment: 70 lbf/ft² = 3.50 kN/m²
                   factory floors: 100 lbf/ft² = 5.00 kN/m²

Safe bearing capacity of soil

1 ton/ft²   = 107.25 kN/m²
2 tonf/ft²   = 214.50 kN/m²
4 tonf/ft²   = 429.00 kN/m²

Stresses	100 lbf/in.² = 0.70 MN/m²
in	1000 lbf/in.² = 7.00 MN/m²
concrete	3000 lbf/in.² = 21.00 MN/m²
	6000 lbf/in.² = 41.00 MN/m²

Costs

$$£1/m^2 = £0.092/ft^2$$
$$1 \text{ shilling (5p)}/ft^2 = £0.538/m^2$$
$$£1/ft^2 = £10.764/m^2 \text{ (approximately } £11/m^2)$$
$$£2.50/ft^2 = £27/m^2$$
$$£5/ft^2 = £54/m^2$$
$$£7.50/ft^2 = £81/m^2$$
$$£10/ft^2 = £108/m^2$$
$$£15/ft^2 = £162/m^2$$
$$£20/ft^2 = £216/m^2$$
$$£25/ft^2 = £269/m^2$$
$$£30/ft^2 = £323/m^2$$
$$£40/ft^2 = £431/m^2$$
$$£50/ft^2 = £538/m^2$$

Index

brick walls   157, 159
concrete floor   156, 158
concrete roof   160
cover   159
damp-proof course   160
door   161–2
excavation for   155
window   163
Pumping   68, 80, 246
chamber   112–19
plant   68

Quadrants   206
Quantities   1, 38, 42, 55
principal   49–50
Quantum meruit   12
Quay
decking   174–5
fender piles   175–6
piling   172–4, 175–6
Query sheets   78

Rails   237, 238, 240, 242
Railway track   238–42
bridge decks   107
buffer stops   238, 241
diamond crossings   237, 241
laying   237, 238, 241
specification   238–9
supply of materials   227, 240, 241
taking up track   238, 242
turnouts   238, 241
Ranges of dimensions   53
Reinforced concrete   104–5, 115–16
pumping chamber   112–19
road   195–6, 204–5
Reinforcement   46, 104–5, 118–19,
161, 174
Reinstatement   211, 224–5
Renovation
sewers   243–6, 247–50
water mains   246, 251
Rescission   12
Reservoir   49–50
Retention money   12–13, 33–4
Rivets   188
Roads   195–208
channels   196, 205
concrete   204–5, 206
edging   196, 206, 207
excavation for   200–4
expansion joints   196, 205
granular base   195, 204, 207

gullies   208
kerbs   196, 205, 206
macadam   195, 207
quadrants   206
setts   206
Rock   50–1, 87, 88, 89, 176
Rock-filled gabions   178, 179
Rogue items   259
Rubbing pieces   182–3
Running sand   88

Samples   80–1
Sand drains   82
Schedules
contracts   15–16
manhole   218–19
of dayworks   26, 28, 51, 70
sewer   216–17
Sea wall
concrete   149–50
excavation   147–9
formwork   150–1
masonry   152
paving to promenade   153
steel sheet piling   149
stone faced   146–53
Seeding   89, 99, 203–4
Selective tendering   41
Setts   206
Sewage treatment works   39–40
Sewers
annulus grouting   245, 249
cleaning   244–5, 247
closed circuit TV surveys   244, 247
external grouting   244, 248
in tunnel   234–6
interruptions   246, 250
junctions   210, 223
local internal repairs   245, 247
markers   225
plugging laterals   245, 247
removing intrusions   245, 247
renovation   243–6, 247–50
schedule   216–17
segmental linings   232, 235, 243,
249
stabilisation   243, 245
stoppers   223
Sills   165
Silting   88
Simpson's rule   90–2
Site accommodation   67, 69
Site clearance   82–6